INTRACELL
PROTEI
DEGRADATION

IN FOCUS®

Titles published in the series:

***Antigen-presenting Cells**
***Complement**
DNA Replication
Enzyme Kinetics
Gene Structure and Transcription 2nd edn
Genetic Engineering
***Immune Recognition**
Intracellular Protein Degradation
***B Lymphocytes**
***Lymphokines**
Membrane Structure and Function
Molecular Basis of Inherited Disease 2nd edn
Molecular Genetic Ecology
Protein Biosynthesis
Protein Engineering
Protein Targeting and Secretion
Regulation of Enzyme Activity
***The Thymus**

***Published in association with the British Society for Immunology.**

Series editors

David Rickwood

Department of Biology, University of Essex, Wivenhoe Park,
Colchester, Essex CO4 3SQ, UK

David Male

Institute of Psychiatry, De Crespigny Park, Denmark Hill,
London SE5 8AF, UK

INTRACELLULAR PROTEIN DEGRADATION

Fergus J. Doherty and
R. John Mayer

Department of Biochemistry, University of Nottingham Medical School,
Queen's Medical Centre, Nottingham NG7 2UT, UK

IRL PRESS
at
OXFORD UNIVERSITY PRESS

Oxford University Press
Walton Street, Oxford OX2 6DP

Oxford is a trade mark of Oxford University Press

Published in the United States
by Oxford University Press, New York

British Library Cataloguing in Publication Data
A catalogue record for this book is available from the British Library

Library of Congress Cataloging in Publication Data
Doherty, Fergus J.
Intracellular protein degradation/Fergus J. Doherty and
R. John Mayer
p. cm. – (In Focus)
Includes bibliographical references and index.
1. Proteins – Metabolism. 2. Cell metabolism. I. Mayer, R.J.
II. Title. III. Series: In Focus (Oxford, England)
[DNLM: 1. Cells – metabolism. 2. Proteins – metabolism.
QU 55 D655 i]
QP551.D62 1992 574.87'61 – dc20 92 – 1432
ISBN 0-19-963293 – 6

In Focus is a registered trade mark of the Chancellor, Masters, and Scholars of the University of Oxford trading as Oxford University Press.

Typeset and printed by Information Press Ltd, Oxford, England.

To Lynne, Kieran, and Conn
To Elaine, Nicholas, and Timothy

Preface

The turnover of intracellular proteins is of fundamental biological importance but has been a relatively neglected and poorly understood area until recently. We have tried to present an overview of intracellular protein degradation while at the same time introducing some of the latest findings which illuminate the underlying molecular mechanisms.

Fergus J. Doherty
R. John Mayer

Contents

Abbreviations and symbols

ATP	adenosine triphosphate
$d[P]/dt$	rate of change of protein concentration
$d[R]/dt$	rate of change of concentration of radiolabelled tracer
DHFR	dihydrofolate reductase
E_1	ubiquitin activating enzyme
E_2	ubiquitin ligase
E_3	ubiquitin 'recognin'
FBPase	fructose 1,6-bisphosphatase
HMG-CoA	hydroxymethylglutaryl coenzyme A
kDa	kilodaltons
k_d	first-order rate constant for protein degradation
k_s	zero-order rate constant for protein synthesis
MPC	multicatalytic protease complex
ODC	ornithine decarboxylase
Pfr	far-red absorbing form of phytochrome
Pr	phytochrome
RuBisCO	ribulose 1,5-bisphosphate carboxylase
$t_{1/2}$	protein half-life

Abbreviations for amino acids

Amino acid	Single letter code	Triple letter code
Alanine	A	Ala
Arginine	R	Arg
Asparagine	N	Asn
Aspartic acid	D	Asp
Cysteine	C	Cys
Glutamic acid	E	Glu
Glutamine	Q	Gln
Glycine	G	Gly
Histidine	H	His
Isoleucine	I	Ile
Leucine	L	Leu
Lysine	K	Lys
Methionine	M	Met
Phenylalanine	F	Phe
Proline	P	Pro
Serine	S	Ser
Threonine	T	Thr
Tryptophan	W	Trp
Tyrosine	Y	Tyr
Valine	V	Val

1

The protein turnover 'cycle'

1. Introduction

Biological molecules are not only continuously synthesized but they are also degraded. However, the notion that biopolymers, for example RNA and proteins, are continuously synthesized and degraded was not considered seriously until the introduction of stable isotopes and the monograph of Schoenheimer *The dynamic state of body constituents*. It should be noted that although DNA is synthesized before mitosis and meiosis, it is not thought to be continuously synthesized and degraded. If these biopolymers, proteins and RNA, are both synthesized and degraded to amino acids and nucleotides respectively, then there must be regulated biochemical mechanisms involved in both the synthesis and degradation of each ribonucleic acid and protein molecule. This book is aimed as an introduction to the mechanisms and regulation of intracellular protein degradation.

In contrast to the progress in understanding protein synthesis, the biochemical mechanisms involved in the degradation of RNA and proteins are not so well understood. Perhaps human nature dictates that there is much more aesthetic pleasure in constructing edifices than in destroying them! This philosophy was the driving force behind the successful search for the mechanisms of RNA and protein synthesis. There is even still a feeling amongst some biological scientists that biopolymer degradation is at best a curiosity and at worst a nuisance. In addition, the measurement of the rates of degradation of proteins has never been an easy task. However, it is now realized that the mechanisms involved in the degradation of RNA and proteins are just as important in cellular homeostasis as biosynthetic mechanisms, and by controlling the concentration of regulatory proteins degradation has been identified as one of the primary control points of some biological processes.

2. Kinetics of protein turnover

Proteolysis is widespread in cells and can be divided roughly into two types: namely, limited proteolytic events, for example to generate mature forms from polypeptide precursors, and secondly complete breakdown of proteins to amino acids. It is with the latter process that this book is concerned, and how protein degradation is defined. Cells and tissue hypertrophy (i.e. increase in the protein content of cells), as well as cellular and tissue atrophy (i.e. loss of proteins from cells), is achieved by the differences in the rates of synthesis and degradation of proteins. The steady-state concentrations of each enzyme and protein in a cell are similarly regulated by the balance between the rate of synthesis and degradation of each protein. The theoretical and experimental reasons for these statements will be analysed to explain the kinetic findings on which they are based.

Proteins are turned over with zero-order rates of synthesis and first-order rates of degradation. What does this mean in biochemical and cellular terms? The genetic code specifies the position in a protein sequence of every amino acid. During synthesis the amino acid requirement at a single position is dictated by the appropriate codon in the mRNA. This obligatory requirement must be met. If the amino acid is not available, then protein synthesis must abort and the truncated polypeptide will detach from the ribosome and be degraded (translation-coupled degradation of polypeptides is but one example of an intracellular protein degradation system; see Section 2 of Chapter 4). In addition, no proteins will be produced without the appropriate mRNAs, tRNAs, ribosomes, and the other factors necessary for synthesis to occur. This is what is known as zero-order kinetics. The protein synthesizing apparatus must be saturated with the factors and macromolecules needed to make proteins, otherwise protein synthesis will not take place.

The kinetics of intracellular protein degradation is not zero-order, it is first-order. By first-order is meant that the reaction rate is dependent and limited by the concentration of the reactants. The rate of degradation of a protein can be measured by the pulse – chase approach (Chapter 2). For the simplest situation of cells in culture a pulse with an appropriate amino acid tracer (for example, [^{3}H]-leucine or [^{35}S]-methionine) for a short period (for example, 15 – 30 min) introduces a label into proteins by incorporation of the amino-acid tracer during protein synthesis. Loss of radiolabelled amino acid from the proteins occurs because of degradation of the proteins to amino acids (but see Chapter 2). In most cases the loss of labelled amino acid is exponential, or is taken to be exponential (*Figure 1.1*). This is the observation for the disappearance of a substance with first-order kinetics during a reaction process and is expressed by equation [1.1], where $[R]_0$ is the initial radiolabelled protein concentration, $[R]_t$ the radiolabelled protein remaining after time t, and k_d is the fractional rate constant for degradation and has the units t^{-1} (hours^{-1}, days^{-1}, etc.).

$$d[R]_0/\mathrm{d}t = -k_\mathrm{d}[R]_t \qquad [1.1]$$

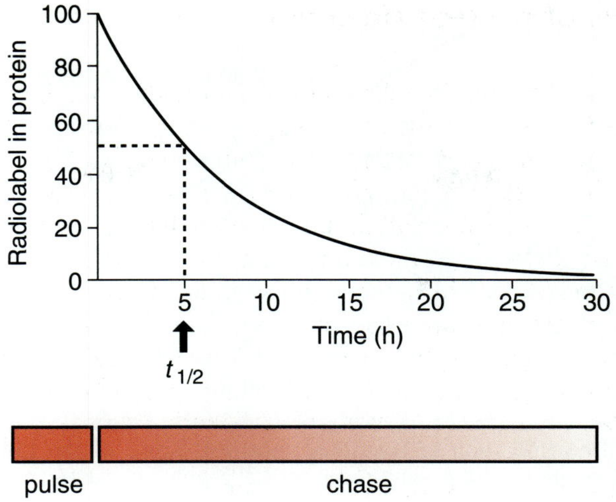

Figure 1.1. Pulse–chase radiochemical analysis to determine the half-life of an intracellular protein. Cells are labelled for a short time (pulse) with a radiolabelled amino acid, washed several times in medium containing the non-radioactive version of the amino acid, and cultured for a prolonged period in the presence of the non-radiolabelled amino-acid to study protein degradation (chase). At intervals the cells are homogenized and the protein(s) of interest purified (for example, by immunoprecipitation). The radioactivity in the protein is plotted against time. The half-life is given when the radioactivity in the protein has fallen by half; 5 h in this example.

integrating and taking natural logarithms:

$$ln\ [R]_t = ln\ [R]_0 - k_d t \qquad [1.2]$$

rearranging:

$$k_d = (ln[R]_0 - ln\ [R]_t)/t \qquad [1.3a]$$

$$k_d = ln\ ([R]_0/[R]_t)/t \qquad [1.3b]$$

A further consequence of this type of loss of radiolabel from the proteins, as with the radioactive decay of isotopes, is that by taking logarithms to the base 10 the following equation is obtained:

$$\log_{10}\ [R]_0 = \log_{10}\ [R]_t - 2.303\ k_d t \qquad [1.3c]$$

Expressing $\log_{10}[R]_t$ against time gives a straight-line plot from which the slope ($\times 2.303$) gives a first-order rate constant of degradation (k_d) of the protein (*Figure 1.2*). A convenient way of expressing first-order rate constants is as half-lives. The half-life is the time taken for loss of half the protein molecules; that is, when $[R]_t$ has fallen to half the initial amount ($[R]_0$), then $[R]_0/[R]_t$ is equal to two, and the time taken (half-life, $t_{½}$) is given by rearranging [1.3b]:

$$t_{½} = ln\ 2/k_d \qquad [1.4]$$

$$t_{½} = 0.693/k_d \qquad [1.5]$$

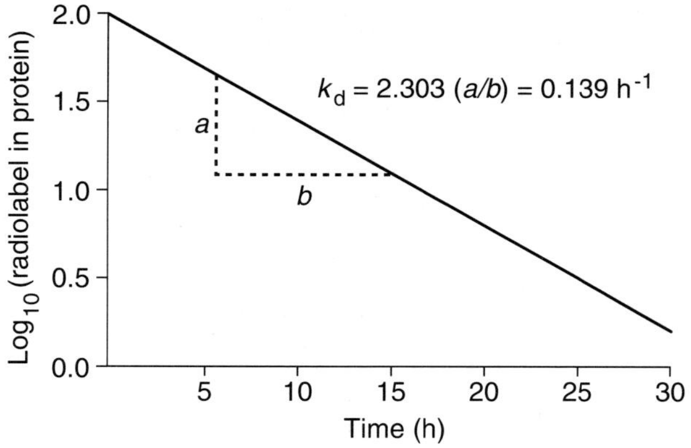

Figure 1.2. Determination of fractional rate (k_d). Log_{10} of the radioactivity in the protein is plotted against time. The slope × 2.303 gives k_d. Data from *Figure 1.1* give a k_d of 0.139 h^{-1}.

In this way, each protein can be considered to have a half-life in the cell. The concept of half-life has the advantage of ease of use and accepted meaning. Generally, in mammalian cells housekeeping, cellular proteins have half-lives of 24 – 72 h (*Table 1.1*). However, regulatory enzymes and proteins (for example, transcription factors) have very short half-lives of the order of 5 – 120 min (*Table 1.1*).

The regulatory significance of these differences will be discussed later in this section. An alternative way of thinking of degradation rate constants is to convert them into percentage rates; that is:

$$\text{Protein degradation } (\% \, t^{-1}) = k_d \times 100 \qquad [1.6]$$

It is now possible to think of the percentage of a protein degraded per hour, or per day. Rapidly turning over proteins (for example, regulatory proteins) will be degraded at, say, 10 – 100% per hour, whereas slowly turning over proteins will be degraded at less than 10% per hour. Another interesting way of thinking of fractional rates of degradation in mature cells in a steady state is as fractional rates of replacement of a protein in the cell; that is, a protein is replaced at 1% per day or 100% per day. Why should this be so?

To understand this notion, it is necessary to consider the consequences of a zero rate of protein synthesis and a first-order rate of degradation on the rate of change in the amount of a protein in a cell, and having established this relationship to consider the consequences for the steady state.

The rate of change in the amount of a protein can be expressed as:

$$d[P]/dt = k_s - k_d[P] \qquad [1.7]$$

where $d[P]/dt$ is the rate of change in the concentration of a protein with time, k_s is the zero-order rate constant of synthesis, k_d is the first-order rate constant of degradation, and $[P]$ is the concentration of the protein. Clearly, integration

Table 1.1 Some intracellular protein half-lives in mammalian cells

Protein	Subcellular site for function	Half-life ($t_{1/2}$)
'Household' enzymes		
Acetyl-CoA carboxylase	Cytosol	48 h
Arginase	Cytosol	96 h
Alanine aminotransferase	Mitochondrion	18 h
Catalase	Peroxisome	60 h
Cytochrome P450	Endoplasmic reticulum	50 h
Regulatory enzymes		
Ornithine decarboxylase	Cytosol	10–30 min
Fructose 1,6-bisphosphatase	Cytosol	30 min
Hydroxymethyl CoA reductase	Endoplasmic reticulum	120 min

of this equation will allow the calculation of the concentration of the protein at any time after some perturbation of a steady state. However, the simplest consequence of the relationship is:

when

$$d[P]/dt = 0 \quad [1.8]$$

then

$$0 = k_s - k_d[P] \quad [1.9]$$

which rearranges to give

$$[P] = k_s/k_d \quad [1.10]$$

The equation shows that in the steady state the concentration of every protein in a cell is determined by the ratio of its rates of synthesis and degradation. Therefore, the fractional rate of protein synthesis in a steady state is equal to the fractional rate of degradation of the protein.

The kinetic relationship between protein synthesis and degradation described above has great implications for the chronic regulation of the amounts of proteins. This can be best illustrated by considering the effect of protein half-life on the steady-state pool size of any two proteins, A and B. (*Table 1.2*). Two proteins A and B have the same rates of synthesis, but the degradation rate of A is 100-fold faster (*Table 1.2*). Use of equation [1.10] predicts that in the steady state the concentrations of A will be 100-fold lower. If the synthesis rates of both proteins A and B are increased tenfold in response to some stimulus, the integrated form of the rate equation [1.7] predicts that after just one hour there will an increase of approximately sevenfold in the concentration of protein A, whereas there will only be a 1.09-fold change in the concentration of protein B. Clearly, the faster turnover and relatively small pool size of protein A means that the change in the concentration of protein A in response to the change in synthetic rate is much faster and more dramatic than the change in the concentration of protein B.

Table 1.2 Protein degradation rates of regulatory and non-regulatory proteins

Protein	$t_{1/2}$ (h)	k_d (h^{-1})	k_s	Enzyme concentration (units)	Fold increase
Steady state					
A	0.693	1.00	1	1	–
B	69.3	0.01	1	100	–
Non-steady state					
A	0.693	1.00	10	6.7	6.7
B	69.3	0.01	10	109	1.09

Two proteins A and B have the same rates of synthesis (k_s) but rates of degradation (k_d) which differ 100-fold. In the steady state, the concentrations of the proteins differ 100 times. In the non-steady state after 1 h with a tenfold increase in the rate of synthesis of both proteins the concentration of protein A has increased nearly seven times, while that of protein B has hardly changed.

Protein A is likely to be a regulatory protein involved in some metabolic or growth control pathway because it is present in low concentration and because its concentration can be changed very rapidly following a stimulus.

The concentration of the rapidly degraded protein A can be decreased as quickly as it is increased; for example, if the synthesis of the protein is switched off. The degradation rate of a protein is as important as its synthesis rate in determining the concentration of the protein in the cell. The simple model in *Table 1.2* only considers a change in the rate of protein synthesis and examines the influence of protein degradation rate on the change in the concentration of a protein. However, there is no reason why only the synthesis rate of a protein should change. There are well-documented instances where only the degradation rate changes, or the more complex situation where both the rates of synthesis and degradation of a protein may change in response to a stimulus.

The importance of degradation rate as a major regulator of cellular homeostasis is illustrated by several proteins. The cell cycle is controlled by the stability of the protein cyclin (Chapter 3, Section 3.2). This protein is degraded in a cyclic manner. The destruction of cyclin triggers the commencement of mitosis. Similarly, phytochrome, which controls many light-sensitive processes in plants, is degraded in a light-sensitive manner, resulting in major changes in photosensitive processes (Chapter 3, Section 3.2). The enzyme, hydroxymethylglutaryl coenzyme A (HMG-CoA) reductase, which regulates the biosynthesis of cholesterol, has a very short half-life (2–4 h) and exhibits cyclic variation in concentration which is controlled by changes in both the rate of synthesis and degradation of the enzyme (Chapter 5, Section 5).

The theoretical inference from exponential loss of radiolabel from protein is that protein degradation is random. A newly synthesized protein is just as likely to be degraded as a protein which has been synthesized for some time. There is no evidence that proteins show lifetime kinetics (that is, they 'wear out') although 'damaged' proteins are rapidly removed by degradation. However, lifetime kinetics are shown for the elimination of red blood cells *in vivo*, and therefore one of the few proteins to show lifetime kinetics for degradation is haemoglobin.

3. Cell biology and a protein turnover cycle

The synthesis and degradation of proteins take place in the context of the complex three-dimensional structure of the cellular architecture. The continuous production and degradation of proteins should be viewed as a protein turnover cycle (1) (*Figure 1.3*). Proteins which are synthesized on free or membrane-bound ribosomes are dispatched to sites (organelles) where they function. Within the primary sequence of each protein is targeting sequence information which directs each protein to some functioning site (for example, nucleus, mitochondrion, endoplasmic reticulum lumen, lysosomes, endosomes, and cell membrane). Proteins, synthesized on free ribosomes without such sequence information, presumably default to the cytosolic space. These processes of protein production and targeting collectively constitute the anterograde arm of the protein turnover cycle in eukaryotic cells (*Figure 1.3*).

Given that proteins are degraded, where does this take place? There is some evidence that proteins can be degraded in mitochondria, chloroplasts, lumen of the endoplasmic reticulum, and endosomes (see Chapter 3). However, the two major characterized systems of intracellular proteolysis are found in the cytosolic space and in the lysosomes. Proteins resident in any organelle may presumably be degraded in that organelle, or leave the organelle for degradation in the cytosolic space, or in the lysosomal system. There is good evidence that proteins

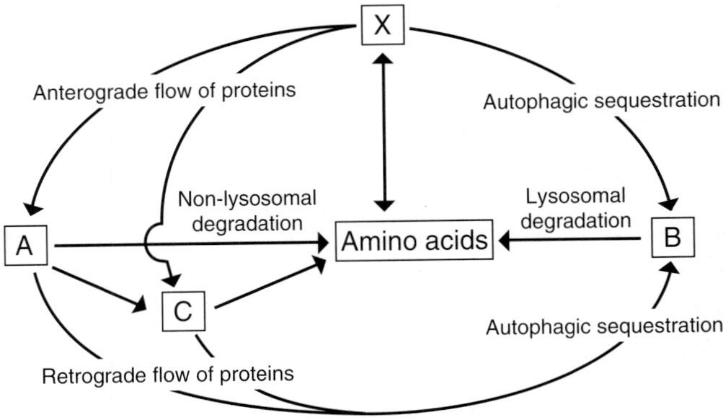

Figure 1.3. Protein turnover cycle in eukaryotic cells. Proteins synthesized on free or membrane-bound ribosomes are dispatched to functioning sites by anterograde flow. Subsequently, proteins are either degraded *in situ* or leave functioning organelles and move by retrograde flow to sites of protein degradation. Newly synthesized proteins (X), protein functioning sites (organelles) in the cell (A), cytosol (C), and sequestration site for the autophagic pathways (B).

in intracellular membranes and in the plasma membrane can be degraded in the lysosomal system, as well as evidence that nuclear proteins can be degraded by extranuclear non-lysosomal and lysosomal mechanisms. Therefore, there is a retrograde arm of a protein turnover cycle for some proteins (*Figure 1.3*) in which proteins may leave their functioning sites for degradation at one of the two major sites of intracellular proteolysis. Such a cycle must explain the fact that in mature cells in a steady state, proteins are degraded with a variety of half-lives (*Table 1.1*). Thus, proteins in a single organelle are degraded at different rates. Therefore the proteins must be differentially sensitive to intra-organellar protein degradation, or leave organelles at different rates for degradation by the major cellular proteolytic systems (*Figure 1.3*). The heterogeneity of the degradation rates of proteins in a single organelle is quite limited, which implies that there may be relatively small numbers of groups (families) of proteins in each organelle with similar rates of degradation and synthesis. Proteins in each family should have sequence motifs which determine the mechanism and rate of their destruction; for example, whether they can leave their functional site for degradation or be degraded *in situ*. There is a further possibility that the enzymes and ancillary factors of a cytosolic proteolytic system may enter an organelle to degrade proteins.

4. Further reading

Schoenheimer, R. (1942) *The dynamic state of body constituents* (E.K. Dunham Lectures, 1941). Harvard University Press; reprinted Hafner, New York (1964).

Waterlow, J.C., Garlick, P.J. *et al.* (1978) *Protein turnover in mammalian tissues and in the whole body*. North-Holland, Amsterdam.

Waterlow, J.C. and Stephen, J.M.L. (1981) *Nitrogen metabolism in man*. Applied Science Publishers, London.

5. Reference

1. Mayer, R.J. and Doherty, F.J. (1986) *FEBS Letts,* **198,** 181.

2

Methods used to study intracellular proteolysis

1. Cells in culture and isolated organs

Whole organisms, laboratory animals and human subjects, isolated organs and cells, and cell-free systems have been used to study intracellular protein degradation. Whole-animal studies are particularly important for an understanding of the response of protein turnover to complex changes such as starvation and disease. An understanding of the molecular events in protein degradation has come primarily from studies on isolated cells in culture and to a lesser extent from isolated organs. Only recently have cell-free systems been developed.

1.1 Pulse–chase methodology

Protein degradation can be determined by measuring the decline in protein concentration (for example, of an enzyme by measuring its activity) if protein molecules that are degraded are not replaced by synthesis. With ongoing protein synthesis replacing proteins that are destroyed, changes in protein mass will not be indicative of the extent of protein breakdown. Blocking protein synthesis with inhibitors (such as cycloheximide) may affect the processes involved in protein degradation, therefore protein degradation is often measured by following the loss of radiolabelled protein from a pool of pre-labelled protein molecules. Precautions must be taken to minimize reutilization by protein synthesis of the label released by protein breakdown. This is most easily accomplished in cells in culture. Following an initial *pulse* with radiolabelled tracer amino acid precursor, cells are washed free of radiolabelled amino acid and radioactivity *chased* out of protein by degradation in the presence of an excess of the unlabelled form of the tracer. The high concentration of the unlabelled tracer during the chase dilutes the radiolabelled amino acid released by protein breakdown and minimizes its reutilization for protein synthesis. Successive measurements of the label left in protein during the chase period permit the calculation of the rate constant of degradation (k_d, see Chapter 1, *Figures 1.1* and *1.2*).

As the degree of incorporation of label during the pulse reflects the rate of turnover of proteins, short periods of pulse labelling, of the order of minutes, allows the monitoring of the degradation of the population of short-lived proteins largely in the absence of long-lived protein degradation. Long pulse periods, 6 h and more, lead to a preferential incorporation of label into long-lived proteins, and the loss of radioactivity from protein will reflect the turnover of long-lived proteins. In addition, inclusion of inhibitors of protein degradation during the chase period may aid in identifying the contribution of different proteases and proteolytic pathways to turnover.

The degradation of individual proteins can be estimated by pulse labelling and determining the radioactivity remaining in the individual protein during the chase period. This can be accomplished by estimation of the label in a protein precipitated by a specific antibody. Complex mixtures of radioactive proteins may also be resolved by gel electrophoresis under denaturing conditions, and the radioactivity in individual polypeptide species estimated. The high resolution of two-dimensional electrophoresis is particularly useful in this context. Radiolabelled prosthetic groups are also used to label specific proteins (for example, ^{3}H-labelled biotin in the carboxylases).

1.2 Gene manipulation

Examination of the relationship between a protein's sequence and structure and its intracellular proteolytic stability requires manipulation of the protein sequence. This can be done by changes to the information coding proteins, or by direct modifications of the proteins themselves. Changes in protein sequence can be effected by manipulation of genes which can then be introduced into the degradative environment of a recipient cell where the fate of the expressed protein can be followed by pulse–chase methods (1). Proteins of interest can also be 'tagged' by generation of chimeric genes encoding the test protein, or a fragment of the test protein, fused to a reporter protein, most notably the readily detectable β-galactosidase. The influence of the identity of the amino terminal residue on degradation by a cytosolic degradative system was identified in this way (Chapter 4, Section 3). Alternatively, the degradative mechanisms of a cell may be probed by genetic methods.

1.3 Microinjection

Proteins may be altered by chemical modification or limited proteolysis in the test-tube. The effect of these changes on their proteolytic stability can be determined following introduction into recipient cells by microinjection. The microinjection of large numbers of culture cells can be achieved by the use of red blood cells as 'microsyringes' (2). In this procedure, erythrocytes, made permeable to macromolecules by swelling in a hypotonic solution of the labelled protein of interest, are resealed by a return to isotonic solution, trapping a fraction of the labelled protein within the red-cell lumen (*Figure 2.1*). Fusion of the red cell to plasma membrane of a recipient cell with the aid of chemical or viral fusogens introduces the labelled protein into the cytoplasm of the recipient cell.

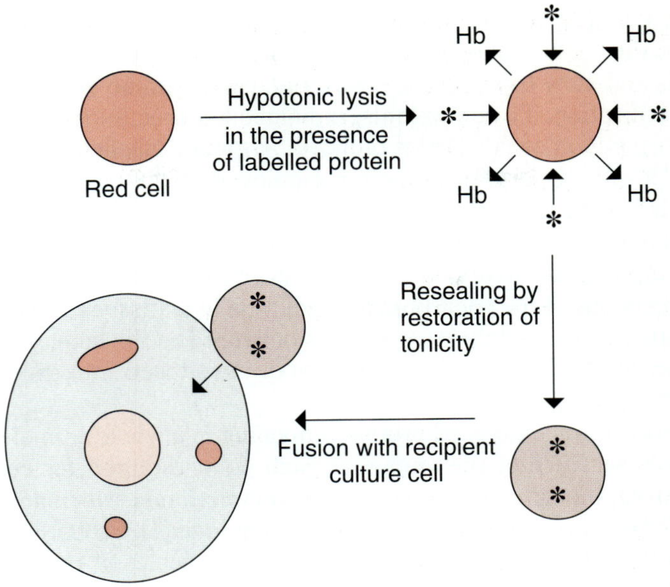

Figure 2.1. Erythrocyte-mediated microinjection of radiolabelled proteins into culture cells. Erythrocytes are lysed in a hypotonic solution of the radiolabelled protein of interest (∗). Protein entering the lumen of the erythrocyte is trapped on subsequent restoration of tonicity and membrane resealing. 'Loaded' erythrocytes deliver radiolabelled protein into the cytoplasm of recipient culture cells following fusion of culture cell and erythrocyte plasma membrane with chemical fusagen (polyethylene glycol) or fusagenic virus (Sendai). This technique has enabled rate and site of degradation of the introduced protein to be determined.

Microinjection of a range of proteins, pre-radiolabelled *in vitro* with ^{125}I-iodine, revealed the pentapeptide motif responsible for targeting some proteins for lysosomal degradation (Chapter 4, Section 5). Microinjection into the cytosol of non-protein radiolabelled probes, such as the membrane impermeant raffinose, sucrose, and lactose, by electropermeabilization of the plasma membrane, has enabled estimations of the non-specific autophagic sequestration of cytoplasmic constituents and the interrelationships between various vesicular cell compartments involved in protein degradation (3) (Chapter 3, Section 2).

2. Whole animals

2.1 Pulse–chase techniques

Pulse–chase techniques are much more difficult to apply in studies involving whole animals. Minimization of reutilization of the label released by protein degradation is difficult in the presence of a number of different pools of radiolabelled precursor (plasma, tissue extracellular and tissue cell free amino acids, tissue charged tRNA) which may not readily equilibrate. Amino acids, released into the tissue following protein catabolism, may readily be reutilized for synthesis and slow to equilibrate with unlabelled amino acid introduced to minimize reutilization. To combat this, radiolabelled precursors may be chosen where reincorporation of the label is reduced. Arginine, labelled in the guanidino group with ^{14}C, [6-^{14}C] arginine, is often employed as the tracer as in the liver the isotope is lost to urea in the urea cycle. Injection of [^{14}C]-$NaHCO_3$ labels in the guanidino carbon of arginine via the urea cycle (in the liver), and the carboxyl group of glutamate and aspartate via carboxylation of pyruvate, and the interconversions of the tricarboxylic acid cycle in all tissues where this cycle is active. A short pulse of label to protein is delivered as the label is rapidly lost by expiration of $^{14}CO_2$. In addition, the label introduced into glutamate and aspartate also rapidly decays.

Determination of radiolabelled protein remaining in a whole animal during the chase involves sacrifice of the animal at each point required for construction of a decay curve, introducing problems due to variation between animals. This problem may be overcome by a double label technique (4). Proteins are labelled by injection of a ^{14}C tracer amino acid and the ^{3}H species of the tracer amino acid is injected at the end of the chase. It is then assumed that the incorporation of ^{3}H is a measure of the initial labelling of proteins and that the $^{3}H/^{14}C$ ratio is a measure of breakdown. This method is also particularly useful in comparing the breakdown of individual proteins resolved by electrophoresis.

2.2 Measurements of amino-acid flux

Measurements of protein-bound radioactivity involve biopsy, or sacrifice of laboratory animals. Non-invasive methods must be employed for measurements of protein turnover in humans (5). For a pool of amino acids available for protein synthesis, the loss of amino acid from this pool will be due to protein synthesis (S), oxidation (E), and other forms of amino acid breakdown (M). Addition to the pool will be a result of breakdown of proteins (B), diet (I), and *de novo* synthesis (N) (*Figure 2.2*). If the free amino-acid pool is constant, then the flux of amino acid through this pool (Q) will be:

$$Q = S + E + M = B + I + N. \qquad [2.1]$$

For an essential amino acid N is zero and M is taken to be insignificant compared to E. Therefore:

$$Q = S + E = B + I. \qquad [2.2]$$

If a labelled tracer amino acid is given as a single injection into a whole animal and the pool of amino acid available for synthesis is taken to be in equilibrium with the plasma pool, then the clearance of label from the plasma is assumed to be indicative of flux of amino acid through the pool (Q) and the loss of amino acids from the pool can be calculated. For leucine, labelled in the carboxyl-carbon oxidation, (E) can be estimated from the expiration of $^{14}CO_2$. Alternatively, ^{15}N-labelled amino acids (glycine, lysine) have been employed and E estimated from urinary ^{15}N (as urea, ammonia). If there is no change in the amino-acid pool size, then breakdown (B) can be calculated from flux minus intake. However,

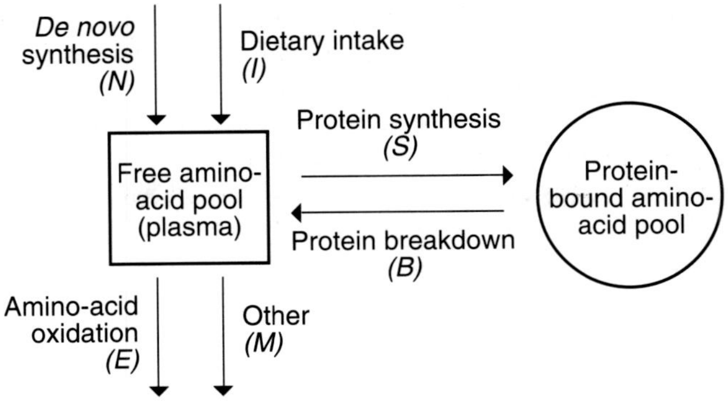

Figure 2.2. Flux of amino acids into and out of the free pool. I = dietary intake, N = *de novo* synthesis, B = amino acids from protein breakdown, S = protein synthesis, E = amino-acid oxidation, M = loss of amino acids by other routes.

if the tracer amino acid is introduced as a constant infusion the specific activity in the plasma reaches a plateau when the introduction of tracer equals its clearance from the plasma. Flux of amino acid can be calculated from the rate of tracer infusion and the specific activity of the tracer at the plateau. Protein breakdown in individual organs can be assessed by the release of essential amino acids from that organ into the blood (i.e. the arterio-venous difference in amino-acid concentration).

2.3 Excretion of modified amino acids

If an amino acid is not utilized for protein synthesis following its release from protein, then its release from the body or tissue would serve as a measure of protein degradation. Actin and myosin, the major proteins of skeletal muscle, contain the post-translationally modified amino acid, *N*-methyl-histidine, which is not reincorporated into protein, or metabolized following release from protein. *N*-methyl-histidine released by protein breakdown is also completely excreted in the urine, therefore *N*-methyl-histidine excretion above the dietary intake is employed as a measure of muscle myofibrillar protein breakdown (6). While the bulk of *N*-methyl-histidine is found in skeletal muscle, there is evidence for a small labile pool in the smooth muscle of the gut and in the skin in some species, which can under some circumstances contribute substantially to the levels of the amino acid in the urine. *N*-methyl-histidine excretion, coupled with urinary creatinine as a measure of muscle mass, is employed as a non-invasive method of estimating muscle protein turnover in man, although this remains controversial (7).

Problems and pitfalls

The methods described have a number of problems. In particular, pulse – chase experiments rely on no reutilization for protein synthesis of the tracer-labelled amino acid released following protein degradation. In practice, this is difficult, if not impossible to achieve, but can be minimized with varying degrees of success. These problems are particularly acute with studies of whole animals, especially man, and doubly so under non-steady-state conditions (negative and positive nitrogen balance) when the protein mass and amino-acid pools of the body are changing. Yet, it is precisely these situations (starvation, trauma, disease) which are of most clinical interest. Researchers have devised many ways to circumvent or minimize these problems which are better dealt with in specialist texts (5).

3. Further reading

Waterlow, J.C. and Stephen, J.M.L. (ed.) (1981)*Nitrogen metabolism in man*. Applied Science Publishers, London.

4. References

1. Bachmair, A., Finley, D. and Varshavsky, A. (1986) *Science*, **234**, 179.
2. Rechsteiner, M. C. (1982) In *Techniques in somatic cell genetics*. (ed. J.W. Shay, p. 385, Plenum Press, New York.
3. Gordon, P.B. and Seglen,P.O. (1988) *Biochem. Biophys. Res. Commun.*, **151**, 40.
4. Arias, I.M., Doyle, D., and Schimke, R.T. (1969) *J. Biol. Chem.*, **244**, 3303.
5. Waterlow, J.C., Garlick, P.J. and Millward, D.J. (ed.). (1978) *Protein turnover in mammalian tissues and in the whole body*. North-Holland, Amsterdam.
6. Ballard, F.J. and Thomas, F.M. (1983) *Clin. Sci.*, **65**, 209.
7. Rennie, M.J. and Millward, D.J. (1983) *Clin. Sci.*, **65**, 217.

3

The mechanisms – pathways of intracellular proteolysis

1. Overview of intracellular proteolysis

The major sites of intracellular protein degradation are the cytosol and the lysosome. Rapidly turning over cytosolic proteins appear to be degraded in the cytosol by proteolytic mechanisms resident there (Section 3), while probably the majority of long-lived proteins are taken up into the dedicated catabolic organelle, the lysosome. Several processes are responsible for the uptake (*sequestration*) of cytoplasmic proteins into the lysosome, a process known as *autophagy* (self-eating, Sections 2.1 and 2.2). The lysosomal system communicates with the extracellular space through endocytic vesicles which are formed at the cell surface and fuse with elements of the lysosomal apparatus following internalization. Consequently, extracellular proteins can also be degraded in the lysosome, a process known as *heteroautophagy* (Section 2.3). Breakdown of proteins to amino acids occurs in cell compartments other than the cytosol (for example, mitochondria and the endoplasmic reticulum; Sections 3.6 and 3.7), which may make a small contribution to overall protein turnover in most cells, but is very important to the regulation of protein concentrations in those organelles.

Intracellular protein degradation is energy dependent, despite peptide bond hydrolysis being exothermic. The requirement for energy, in the form of ATP, is in part due to the ATP-dependent proton pump which maintains the acidic environment of the lysosomal apparatus. However, the formation of autophagosomes is probably more sensitive to intracellular ATP levels. In addition, some macroautophagic post-sequestrational but pre-lysosomal steps (for example, movement of vesicles, fusion) appear to require ATP (1). Under basal conditions (i.e. dividing cells in culture in the presence of amino acids and serum), when non-lysosomal mechanisms are responsible for a major component of total protein turnover, intracellular protein degradation is still largely ATP-dependent. This implies that some cytosolic proteolytic mechanism(s) are ATP-dependent. This is true for the ubiquitin-mediated protein degradation described in Section 3.1.

2. Lysosomal proteolysis

Lysosomes are a family of related cytoplasmic single membrane vesicular structures which contain a variety of hydrolytic enzymes capable of degrading cellular macromolecules (nucleic acids, proteins, polysaccharides, and lipids). A variety of different endoproteases (which cleave internal peptide bonds of proteins) called cathepsins, are found in the lysosome which are classified on the basis of the important amino acid at the active site (*Table 3.1*) and are most active at low pH. The lysosomal membrane contains an ATP-dependent H^+ pump which introduces protons into the lysosomal lumen producing a low lumenal pH of 4.5–5.5. At this pH, the acid proteases of the lysosomes will be most active. Other vesicular cell compartments, such as late endosomes may have a low pH. The lysosome also contains ***exopeptidases*** which 'nibble' from the carboxyl (***carboxypeptidases***), or amino (***aminopeptidases***) termini. In addition, peptidases are present which complete the breakdown of smaller peptides to amino acids. This combination of proteolytic enzymes is therefore capable of the extremely rapid breakdown of the cytoplasmic and extracellular proteins consigned to the lysosome.

The ***primary*** lysosomes (that is, lysosomes which are inactive in degradation) are thought to be formed by a budding process from the Golgi apparatus. Some lysosomal enzymes are directed to the lysosomes by a covalent mannose-6-phosphate recognition signal. A receptor for mannose-6-phosphate-bearing proteins is located in the Golgi membrane and ensures the association of the lysosomal enzymes with vesicles that are directed to the lysososomes. In some cell types, lysosomes may take the form of an extensive anastomosing (branching) tubular system extending from the cell surface to the nuclear membrane. How then do cytoplasmic proteins gain access to the lysosomal space to be degraded? Several mechanisms appear to operate. Lysosomal proteolysis is responsible for bulk protein turnover, and therefore is influenced by the supply of nutrients, especially amino acids and by hormones. This will be dealt with in Chapter 5.

Table 3.1 Lysosomal proteases. In addition, lysosomes contain di- and tripeptidases

Enzyme	Essential active site residue
Cathepsin B	Cysteine
Cathepsin D	Aspartate
Cathepsin G	Serine
Cathepsin H	Cysteine
Cathepsin L	Cysteine
Carboxypeptidase A	
Carboxypeptidase B	
Arginyl aminopeptidase (cathepsin H)	

2.1 Macroautophagy

Autophagy literally means 'self-eating'. There are two mechanisms in cells for autophagy. One mechanism involves wrapping membrane around the cytoplasm to be degraded, and is called *macroautophagy*. The other mechanisms involves the invagination of the membrane of some lysosome-related organelle, and is called *microautophagy*. Both processes may operate in the same cell. In some cell types one mechanism may dominate over the other.

The steps in macroautophagy are depicted in *Figure 3.1*. The first stage in the process appears to be the wrapping of smooth endoplasmic reticulum membrane (which can be derived by degranulation of rough endoplasmic reticulum) around the cytoplasmic material to be degraded to form a double membrane-bound *autophagosome* (2, 3). Whole organelles, such as mitochondria (4), as well as segments of endoplasmic reticulum and cytosol (but not nuclei), can be sequestered in this way. The primary stimulus for the formation of

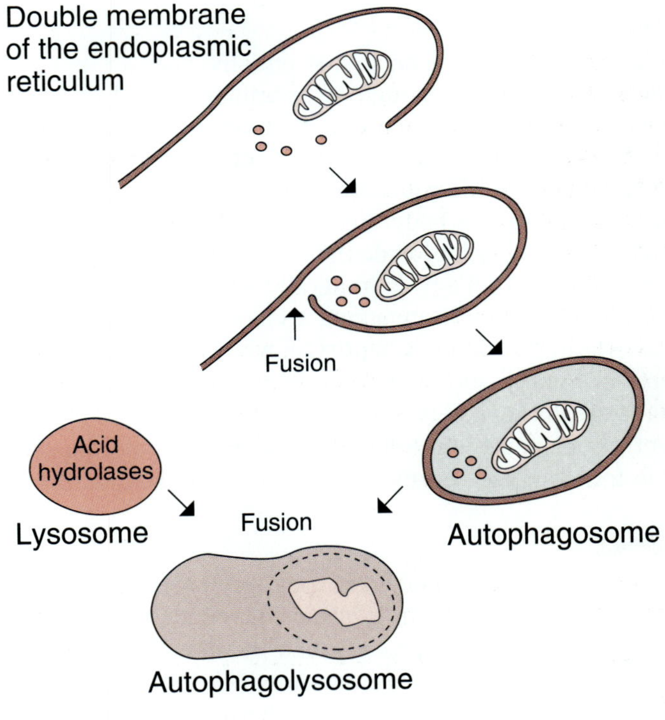

Figure 3.1. Macroautophagy: the entry of proteins into lysosomes by cytoplasmic wrapping by intracellular membrane. A portion of cytoplasm is surrounded and engulfed by a double membrane derived from the smooth (or rough) endoplasmic reticulum. The membrane fuses to produce a double membrane-bound autophagosome. The autophagosome fuses with a primary or secondary lysosome to form an autophagolysosome. The lysosomal hydrolytic enzymes including proteases digest the inner autophagosome membrane and degrade the contents of the autophagolysosome (autolysosome).

autophagosomes is unknown, but the process is energy requiring and presumably needs some focal event in the cytoplasm to trigger the commencement of the wrapping process. Once formed, the autophagosome may fuse with different types of lysosome. Generally, fusion takes place rapidly with a so-called ***primary lysosome*** or with a ***secondary lysosome***; that is, a larger less dense vesicle which has already been involved in previous fusion events with autophagosomes. The autophagosome probably exists for only a few minutes before fusion. As a consequence of fusion, an ***autophagolysosome*** is formed containing all of the acid hydrolases, including proteases, to degrade proteins to amino acids and the proteins themselves destined for destruction. The rate of macroautophagic sequestration can vary enormously with cell type and conditions and is regulated by hormones and nutrients (Chapter 5). Hepatocytes in culture deprived of amino acids can degrade some 5 – 10% of their protein per hour by macroautophagy, while macroautophagy in cultured fibroblasts accounts for generally less than 1% of the total protein per hour.

2.2 Microautophagy

The microautophagic process occurs essentially as described in *Figure 3.2* (5). The lysosome-related vesicle thought to be primarily involved in microautophagy is seen in the electron microscope as a multivesicular body and may be related to the endosome, the distended vesicle to which endocytic vesicles fuse on entry into the cell (*Figure 3.3*). Microinvagination of the surface of the endosome will give rise to a multivesicular body. Fusion of multivesicular bodies with primary or secondary lysosomes will provide the hydrolytic enzymes for macromolecular degradation. Macroautophagy is generally considered to be non-selective in protein degradation, simply consuming areas of cytoplasm for degradation (for example, in liver in starvation; Chapter 5), while microautophagy may be much more selective. Finally, although lysosomal and non-lysosomal proteolysis of intracellular proteins have been seen as separate and unconnected mechanisms, it has recently been demonstrated that a cofactor for non-lysosomal proteolysis, ubiquitin, is found in the lysosomal compartment (Section 3.3).

2.3 Heterophagy

Proteins entering cells by pinocytosis, phagocytosis, or receptor-mediated endocytosis, are found in some form of intracellular endosome-like structure. These distended vesicular ***heterophagosomes*** can fuse with primary or secondary lysosomes to form ***heterophagolysosomes***, in which the material internalized into the cell can be degraded (*Figure 3.3*). The heterophagic and autophagic vesicular systems can communicate, presumably through fusion, as demonstrated by experiments showing that sucrose injected into hepatocytes during electropermeabilization (6) (Chapter 2, Section 1.3) enters autophagosomes and is hydrolysed by sucrase entering the resealed cells from outside by heterophagy. There are too many examples of the importance of heterophagy in cell homeostasis to discuss here, but the process is important for cellular nutrition, hormone and growth factor action (by receptor-mediated endocytosis), and in

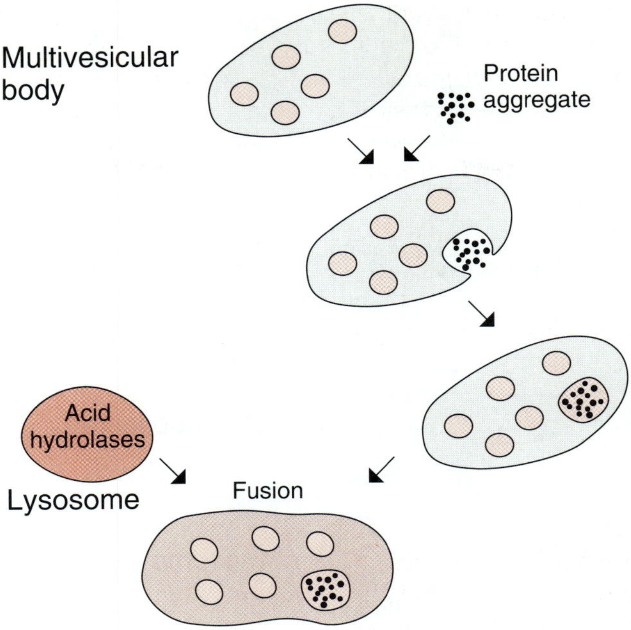

Figure 3.2. Microautophagy: the entry of proteins into lysosomes by the surface invagination of multivesicular bodies. Multivesicular bodies (related to endosomes) invaginate their surface to internalize proteins or protein aggregates. Fusion with primary or secondary lysosomes injects hydrolytic enzymes including proteases into the multivesicular bodies to degrade the contents. Fusion may only occur when numerous vesicles have accumulated inside the bodies. The multivesicular bodies may act like vacuum cleaners internalizing material, perhaps in some selective manner. When 'full', the lysosomal fusion events may take place.

specialized cells such as those involved in the inflammatory system; for example, polymorphonuclear neutrophils and macrophages.

2.4 Secretion-coupled degradation

Many cell types are involved in protein secretion. Proteins synthesized on membrane bound ribosomes are propelled into the lumen of the endoplasmic reticulum, from which, via the Golgi apparatus, they are inserted into secretory vesicles. These then fuse with the cell surface to expel the secretory proteins; for example, the milk protein casein in mammary epithelial cells. This is called the constitutive secretory pathway. Alternatively, some secretory proteins are stored in membrane-bound secretory granules which fuse to the cell membrane to release their protein content when cells receive some extracellular stimulus (for example, insulin and glucagon, which are stored in granules in the neuroendocrine cells of the pancreas). The extracellular signals may be chemical; for

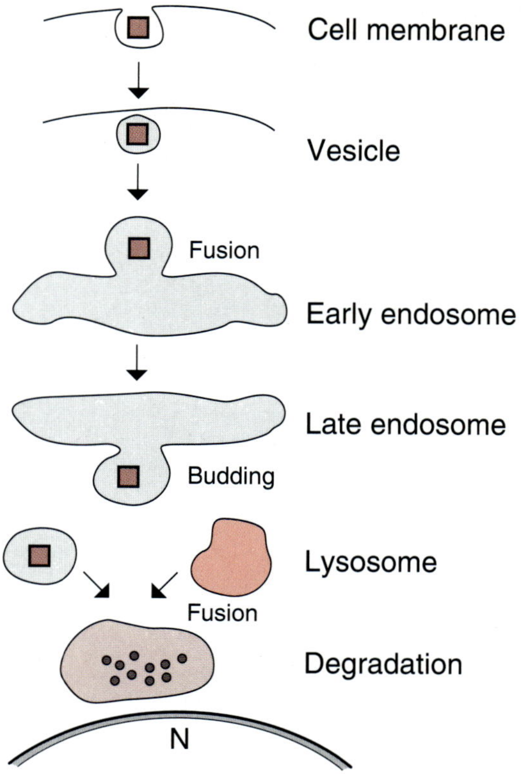

Figure 3.3. Heterophagy: internalization of materials from the outside of the cell into the lysosome-related system. Particles (nutrient polymers) are internalized into the cell by heterophagy. The heterophagosomes move deeper into the cell and fuse with lysosomes which are often arranged around the nucleus (in tissue culture cells) where heterophagolysosomes are formed and the contents degraded by the lysosomal hydrolases including proteases. The cell nucleus is indicated by (N).

example, changes in blood glucose concentration or by nervous stimulation. This secretory pathway is called the regulated pathway.

The secretory rate of the constitutive and regulated pathways can vary. Besides the regulation of the rate of secretory protein synthesis, the degradation rate of the secretory proteins in the membrane-bound vesicles can be varied (7, 8). This is achieved by the fusion of the secretory vesicles, or membrane-bound granules with lysosomes. For example, in the mammary gland the rate of lysosomal degradation of casein in secretory vesicles is inversely proportional to the secretory rate of the mammary epithelial cells, which is dependent on the degree of suckling. The slower the rate of secretion, the faster the rate of lysosome-vesicle fusion and degradation of the casein. In this way, no distention of the mammary epithelial cells occurs due to accumulating vesicles and amino acids from casein can be recycled into new protein synthesis.

3. Non-lysosomal proteolysis

Non-lysosomal protein degradation can be demonstrated in the cytosolic space of cells by soluble protein microinjection (Chapter 2, Section 1.3), by the resistance of a component of protein degradation to inhibitors of lysosomal proteolysis and in reticulocytes which have lost their lysosomal system. Many proteases are found in the cytosol of cells from which they can be purified. Proteolytic inhibitors can be purified from the cytosol of cells which can inhibit neutral proteases and proteases with acidic pH optima. Some of the latter inhibitors may be involved in the inhibition of lysosomal proteases which may enter the cytosolic space by chance, while others may regulate the activity of cytosolic proteases. Given the integrity of the lysosomal membrane, lysosomally mediated protein degradation is only possible following the rate-limiting step of sequestration of substrate proteins from the cytoplasm. Perhaps to protect against the potentially catastrophic effects of damage to the lysosomal membrane, the cytoplasm contains numerous protease inhibitors which will inactivate lysosomal proteases. In addition, the neutral pH of the cytoplasm will tend to inhibit acid protease action. In contrast, non-lysosomal proteases are exposed to possible substrates continually; therefore, cytosolic proteolytic systems must be highly discriminatory if unwanted degradation is not to occur. A complex process of selection and marking of proteins for degradation is the striking feature of the first cytosolic proteolytic mechanism to be considered—that mediated by the small protein ubiquitin.

3.1 The ubiquitin-mediated proteolytic pathway

Ubiquitin is a small globular protein of 76 amino acids with a protruding carboxyl-terminus which was first characterized in the thymus and subsequently described as a ubiquitous cellular peptide (*Figure 3.4*) which targets proteins for degradation (9). The importance of ubiquitin for proteolysis was first realized when it was discovered to be a covalent cofactor for ATP-dependent proteolysis in maturing reticulocytes. These cells have barely detectable lysosomal degradation and yet eliminate proteins and organelles (for example, mitochondria) by an ATP-dependent proteolytic mechanism (10). The proteolytic system can also rapidly degrade abnormal haemoglobin, generated by incubating the cells with amino acid analogues.

Ubiquitin is conjugated to target proteins *via* its carboxyl-terminal glycine residue to form isopeptide bonds with available lysine residues in the target protein (*Figure 3.5*). The result is to produce the novel post-translational modification of branched polypeptides. Target proteins may be multiple ubiquitinated. A chain of up to 20 ubiquitins can be attached to a single lysine residue of a target protein. This is by the addition of ubiquitin units through isopeptide bonds involving lysine-48 of each additional ubiquitin and the carboxyl-terminal glycine residue of each succeeding ubiquitin. Multiple ubiquitination of this type results in recognition of the target protein by proteases (11). If a protein has more than one lysine residue, then multiple ubiquitination can theoretically take place at

Figure 3.4. A space-filling model of ubiquitin showing the compact globular structure. The exposed C-terminal tail and lysine atoms at position 48 are shown in dark grey.

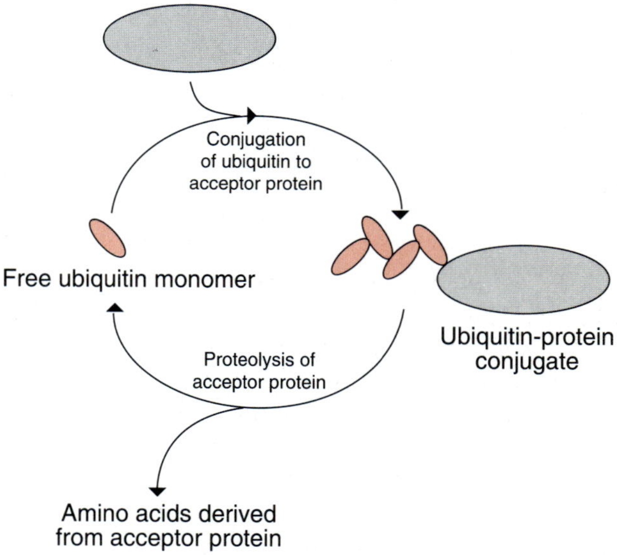

Figure 3.5. A simplified diagrammatic representation of ubiquitin-mediated proteolysis. Tagging with a multi-ubiquitin chain targets the acceptor protein for proteolysis.

each lysine residue, although in fact only one or a few lysines in any one protein act as acceptors for ubiquitin (12). Once a multiple ubiquitin chain has formed, target proteins are rapidly degraded in the cytosol by a large molecular mass protease (Section 3.5). Ubiquitin is released during protein degradation to be used in further rounds of protein catabolism. Ubiquitin can also be removed from proteins by specific ubiquitin carboxyl-terminal hydrolases to regenerate free ubiquitin and the unmodified target protein. This may be part of a quality control system examining each protein continuously and selecting certain protein molecules for degradation while discarding others to continue to function.

Ubiquitin is activated and conjugated to proteins by a complex series of reactions. A general outline of these reactions is shown in *Figure 3.6.* In short, ubiquitin is activated by an activating enzyme E_1 with the hydrolysis of two molecules of ATP to generate two enzyme associated ubiquitins, one as an adenylate and the other linked to the enzyme by a thioester link. The thioester-linked ubiquitin can be transferred to one of a family of ubiquitin-conjugating enzymes, E_2 proteins which can then conjugate ubiquitin to a variety of target proteins. The E_2 family of enzymes is able to conjugate ubiquitin to target proteins for different purposes (13). For example, in yeast, one E_2 protein is identical to a cell cycle regulating protein cdc34. Another E_2 protein in yeast is a radiation repair enzyme RAD6. Two other E_2 proteins in yeast are involved in the conjugation of ubiquitin to the abnormal proteins generated by amino acid analogues before their rapid degradation. The conjugation of ubiquitin to target proteins is also assisted by another enzyme, E_3, which may have a key role in the recognition of the target proteins to be degraded by the ubiquitin-dependent pathway. Ubiquitin undoubtedly has other functions besides targeting proteins for degradation. A fraction of some histone proteins (H2A and H2B) are found to be conjugated to ubiquitin which may regulate gene expression some way. Ubiquitin is also found conjugated to a number of cell surface receptors, the function of this ubiquitination is unknown. Finally, some of the genes encoding for ubiquitin direct the synthesis of fusion proteins where the non-ubiquitin carboxyl-terminal moiety is a ribosomal protein, released following cleavage of the fusion protein to yield ubiquitin and the extension protein (14).

Ubiquitin is encoded by a family of ubiquitin genes which may be expressed differentially in response to different cell stimuli (*Figure 3.7*). There are four genes in yeast, three coding for fusion proteins with carboxyl-terminal extensions involved in ribosomal biogenesis and the other gene coding for polyubiquitin. The polyubiquitin genes in yeast codes for five ubiquitin molecules which are cleaved by a ubiquitin carboxyl-terminal hydrolase into five ubiquitin molecules for use in the cell. Another important feature of the polyubiquitin genes is that they have heat-shock promoters. Therefore, these genes are activated in cells subjected to some form of stress (Section 3.2).

3.2 The physiological role of ubiquitin-mediated proteolysis

Ubiquitin – protein conjugating activity has been demonstrated in muscle, liver, heart, and the eye lens, as well as in yeast and plants, but not in bacteria. It

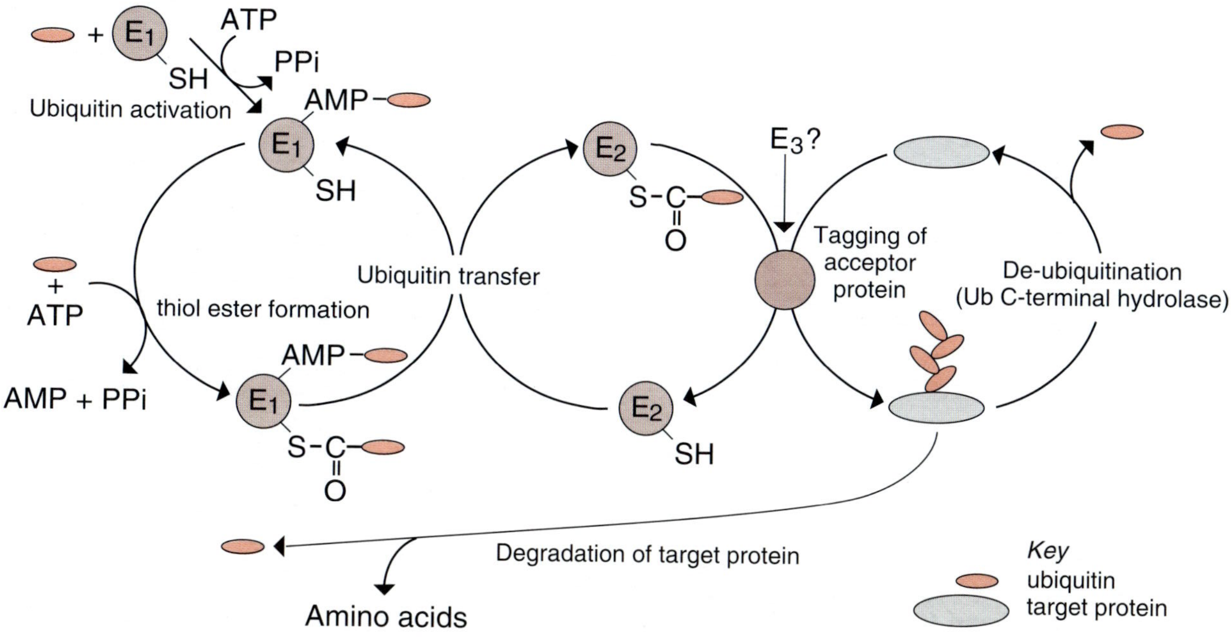

Figure 3.6. The ubiquitin cycle.

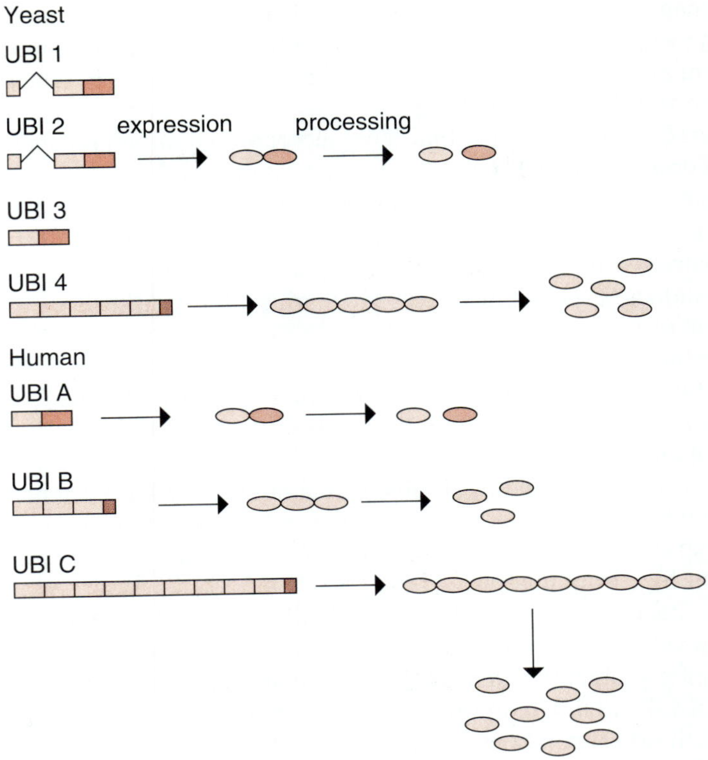

Figure 3.7. Ubiquitin genes in yeast and man. Lines represent introns and boxes exons. Dark orange boxes represent the carboxyl-terminal extensions of the natural ubiquitin fusion proteins which are ribosomal subunits. The polyubiquitin genes code for head to tail repeats of ubiquitin, terminating with an additional amino acid at the carboxyl-terminus of the last ubiquitin. This amino acid is cleaved off by a processing protease which also cleaves the ubiquitin polyprotein products of these genes to ubiquitin monomers.

would appear that ubiquitin-mediated protein degradation plays a key role in eukaryotes. Degradation of short-lived proteins is impaired in mutant cell lines having a temperature-sensitive ubiquitin-activating enzyme E_1 at the non-permissive temperature (39 – 43 °C), suggesting that the ubiquitin system degrades short-lived proteins. One example is the regulation of the intracellular cyclin concentration (17). The transition from metaphase to anaphase is induced by the rapid ubiquitin-mediated degradation of the regulatory protein cyclin. Loss of cyclin inactivates the M-phase promoting factor (MPF, which induces mitosis) and is a key event in the exit from mitosis and entry into interphase. In plants, phytochrome is synthesized as a red-light absorbing form (Pr) and is inter-converted by light irradiation to a far-red-light absorbing form (Pfr). On photoconversion from Pr to Pfr, the half-life of the phytochrome drops from 100 h to 1 – 2 h and ubiquitinated forms of Pfr rapidly appear following illumination which are degraded within a few minutes (18). The ubiquitin-mediated degradation of phytochrome and cyclin are similar in that they demonstrate the role of *selective* protein degradation in controlling intracellular protein concentrations in response to different conditions.

The degradation of proteins that have incorporated an amino acid analogue and are, therefore, judged to be abnormal, is compromised when the ubiquitin-activating activity is inactivated in temperature-sensitive mutants (see above). Ubiquitin-mediated degradation is also responsible for the rapid destruction of proteins damaged by heat. Exposure of organisms to elevated temperatures is known to induce a 'heat-shock' response, where the synthesis of the normal repertoire of proteins is shut down and the synthesis of a limited number of 'heat-shock' proteins' is switched on. This response is thought to be due to the production of heat-damaged macromolecules and macromolecular structures in the cell. Following heat-shock (in this case by exposure to 43 °C) of hepatoma cells, a wave of ubiquitination is observed with numerous high molecular weight ubiquitinated species appearing. Proteolysis of normally stable proteins increases transiently following heat-shock, presumably as some of these are damaged and rapidly degraded (19). Heat-damaged proteins therefore are targeted for degradation by ubiquitination.

Ubiquitin itself is a heat-shock protein, in that elevated temperatures lead to increased levels of ubiquitin following elevated expression of a gene encoding a polyubiquitin protein (Section 3.1), as indeed are the yeast E_2 ubiquitin carrier enzymes UBC4 and 5. Ubiquitin and the enzymes of the ubiquitin pathway act together with other members of the heat-shock protein families to protect the cell from the deleterious effects of heat-damaged proteins. In contrast to the ubiquitin system, the other heat-shock proteins [70 kDa family, 78 kDa (BiP), 80–90 kDa proteins, small heat-shock proteins] may 'protect' vital protein components of the cell; for example, the protein synthetic machinery, from heat-induced damage and consequent degradation.

Ubiquitin-mediated proteolysis has been demonstrated in yeast and plants as well as mammalian cells and tissues, and it is now thought that all eukaryotes contain this proteolytic pathway. Ubiquitin is, in fact, the most highly conserved eukaryotic protein known, which would support its involvement throughout eukaryotes in the same fundamental cellular process.

3.3 Ubiquitin as a unifying factor in intracellular protein degradation

Ubiquitin has not been found in prokaryotes, despite some similarities in cytosolic proteolytic systems between eukaryotes and prokaryotes. Although ubiquitin has been extensively studied as a cofactor for non-lysosomal protein degradation very recent work has suggested a role for ubiquitin in lysosomal proteolysis (14,34), particularly during nutrient deprivation and heat stress (15). The attachment of ubiquitin to proteins may, therefore, be important in both lysosomal and non-lysosomal proteolysis. Indeed, ubiquitin may have evolved as a factor in eukaryotic cell processes, including lysosomal proteolysis, and was then subsequently 'recruited' as a factor for cytosolic protein degradation in eukaryotes. As such, ubiquitin would be a unifying factor in intracellular proteolysis.

3.4 Calpain

Calcium-activated proteolytic activities were detected by studying protease activities which can be activated by micromolar and millimolar calcium ions. Two calpains have been identified which contain two subunits. The smaller subunit (30 kDa) is regulatory and contains an 'EF hand' domain characteristic of some calcium-binding proteins (for example, calmodulin), together with a hydrophobic glycine-rich domain which may associate the enzyme with cell membranes. EF hands are formed by a helix, a loop, and another helix with eight oxygen atoms co-ordinated to each bound calcium. The other larger subunit (80 kDa) also contains an EF hand domain together with a cysteine protease catalytic site which is similar to papain, making the enzyme a potent cysteine protease. The enzymes are obviously regulated by calcium (20). The calpain sensitive to micromolar concentrations of calcium (2 – 75 μM) may be activated in the cytosol, or when bound to the cell membrane or to an activator associated with the cytoskeleton. The form of the enzyme activated by higher concentrations of calcium (0.2 – 0.8 mM) is only activated by autolysis of the membrane form. The smaller 30 kDa subunit is almost identical in both enzymes indicating that the differences in calcium sensitivity lie in the 80 kDa subunit of the enzymes. The enzymes are also regulated by an inhibitor, calpastatin, which binds to calpain at high calcium concentrations. This inhibitor contains four repeats of calpain inhibitory domains.

The activation of calpain by the controlled increase in the concentration of calcium (as a pulse) may result in the hydrolysis of proteins to trigger metabolic change; for example, neutrophil activation. The calpain activity may be controlled in two ways: by calpastatin, and by the subsequent decline in intracellular calcium after some ligand-mediated signal stimulates a membrane receptor. Calpain is implicated in neutrophil and platelet activation (21). Neutrophil activation involves an increase in intracellular calcium, which activates calpain, which in turn proteolytically modifies and activates protein kinase C. The active protein kinase C in turn promotes release of secretory granules following phosphorylation of cytoskeletal components. An intriguing role for calpain has been suggested in long-term potentiation (22), a mechanism that may be a model for memory processes. Stimulation of glutamate receptors increases intraneuronal calcium concentrations and activates calpain, resulting in proteolysis of the membrane cytoskeletal protein fodrin (brain spectrin). Fodrin degradation releases more glutamate receptors to the cell surface and permits remodelling of synaptic morphology. In this way, proteolysis mediates a long-term change (membrane reorganization) induced by a transient event (receptor activation).

Calpain will also degrade intermediate filament cytoskeletal proteins *in vitro* and possibly, *in vivo*. Intermediate filament proteins (vimentin, desmin, cytokeratin, neurofilament proteins, glial fibrillary acidic protein) form stable filamentous structures in cells which may be cleaved by calpain. Neurofilament proteins are protected from calpain attack following phosphorylation. All these examples imply a *regulatory* role for proteolytic attack by calpain to generate possibly large peptide fragments, rather than amino acids. In contrast, the degradation of oxidized haemoglobin by calpain in erythrocytes deficient in glucose-6-phosphate dehydrogenase may delay the formation of large, damaging haemoglobin aggregates (Heinz bodies).

3.5 *Multicatalytic protease complex (MPC)*

The multicatalytic protease, ingensin, or proteasome is a large (20S) protein that behaves as a particle during centrifugal purification and which was described as 'prosome' many years ago (23). The large hollow cylindrical particle (*Figure 3.8*) is a protease quite unlike any other, since it shows catalytic activity against peptide bonds in a variety of peptide sequences. It can cleave on the carboxyl side of basic (trypsin-like), hydrophobic (chymotrypsin-like), or acidic (glutamate) amino acid residues, hence the protease was described as a multicatalytic protease complex (24). On the basis of studies with protease inhibitors, these activities reside on distinct catalytic sites which put the enzyme into a new novel class of proteases. The multicatalytic protease is composed of a large number of different subunits of 21 – 34 kDa and is activated by fatty acids and low concentrations of detergent (sodium dodecyl sulphate), and it has been suggested that normally it exists in a latent form in cells. The complex is found in the nucleus and cytoplasm of cells and may move from one compartment to another when cells are subjected to regulatory or injurious stimuli.

The multicatalytic proteinase may be related to the even larger 26S protease which is involved in the recognition of multi-ubiquitinated protein substrates and which makes the initial endoproteolytic cleavages to begin the degradation of ubiquitinated proteins (25,26). The nature of the cytosolic protease, responsible for the degradation of proteins tagged with ubiquitin, is obviously of great interest. A large molecular weight (1000 kDa, 26S proteasome) protease, named megapain, has been isolated from reticulocytes which degrades lysozyme – ubiquitin conjugates readily in an ATP-dependent process, but free

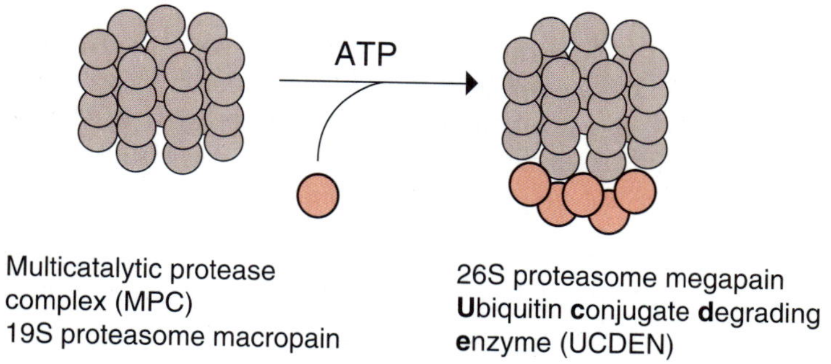

Figure 3.8. The subunit structure of the multicatalytic protease complex. Viewed with the electron microscope the MPC appears to consist of four 6-subunit rings forming a cylindrical particle (19S) with a hollow core. There are at least 10 different subunit types. In the presence of ATP a larger complex forms (26S) with the addition of further subunits which will selectively degrade proteins tagged with a multi-ubiquitin chain.

lysozyme is degraded only poorly. Ubiquitin conjugate-degrading activity is present in the cytosol of skeletal muscle and liver and it has been suggested that it is related to the multicatalytic protease. The multicatalytic protease may form the core of the larger enzyme which is assembled from the multicatalytic core and additional subunits in the presence of ATP. Yeast mutants, lacking multicatalytic protease complex activity, fail to degrade abnormal proteins, respond poorly to stress, and accumulate ubiquitin – protein conjugates. The multicatalytic protease will also readily degrade oxidized proteins and the protease could be responsible for the intracellular degradation of damaged proteins independent of ubiquitin.

3.6 Protein degradation in mitochondria and chloroplasts

Mitochondria and chloroplasts possess a poorly characterized high molecular weight, ATP-dependent but ubiquitin-independent, proteolytic activity (27). This protease may be responsible for the degradation of mitochondrial or chloroplast-encoded mistranslated proteins, damaged mitochondrial proteins, and mitochondrially or cytosolically encoded polypeptide subunits which have failed to combine with their partners to form an oligomer. The plant chloroplast possesses proteolytic activity capable of rapidly degrading the nuclear-encoded subunit of RuBisCO when assembly into the oligomer is prevented by the absence of its chloroplast-encoded partner (28). Both organelles also possess processing proteases responsible for cleaving off N-terminal import sequences from nuclear-encoded mitochondrial proteins.

3.7 Protein degradation in the endoplasmic reticulum

There is a proteolytic activity resident in the lumen of the endoplasmic reticulum which is clearly non-lysosomal, may require ATP, and appears to be selective. Secretory proteins not only have to avoid the process of the secretion-coupled lysosomal protein degradation, which is a quantitative process designed to avoid a build up of secretory vesicles (Section 2.4), but must also undergo a qualitative proteolytic editing carried out by protease(s) in the lumen of the endoplasmic reticulum. Many secretory proteins have multiple subunits which must correctly associate and assemble; for example, undergo disulfide cross-linking in the endoplasmic reticulum. For example, the T-cell receptor complex has seven subunits. Failure of all the subunits to assemble can result in the degradation of the unassembled subunits (29). Immunoglobulins, light and heavy chains synthesized on membrane-bound polysomes, are translocated into the lumen of the endoplasmic reticulum, assembled into oligomers and secreted. In the absence of light chains, heavy chains in the endoplasmic reticulum bind to a member of the heat-shock family of proteins, known as BiPor GRP78, and are degraded (30). As the protease activity of the endoplasmic reticulum is sensitive to the conformation of proteins, correct folding and assembly of polypeptides into oligomers may protect protease-sensitive sites.

Finally, for completeness, as with the mitochondrion there is, of course, a proteolytic system in the membrane of the endoplasmic reticulum which removes hydrophobic leader sequences from incoming proteins and, presumably degrades them to amino acids. The vast majority of proteins entering the endoplasmic reticulum have these signal sequences which need to be degraded. Quantitatively, this proteolytic system degrades a considerable amount of polypeptides in cells which secrete many proteins; for example, hepatocytes and plasma cells.

4. Proteolytic systems in bacteria

Escherichia coli is an intensively investigated organism, but relatively little is known about the mechanisms of protein turnover in this organism and in prokaryotes in general. Bacterial protein degradation most obviously differs from eukaryotes in the lack of an organelle dedicated to degradation—that is, the lysosome. Protein degradation takes places in the cytoplasm and periplasm of *E. coli*. The degradation of abnormal proteins and normal proteins during starvation is energy dependent. There are at least eight soluble endoproteases from *E. coli* one of which, protease La, is ATP-dependent (31). Protease La is the product of the *lon* gene and consists of a tetramer of 93 kDa subunits. Mutations in the *lon* gene reduces the cells ability to degrade abnormal proteins rapidly, including missense and nonsense fragments, mutant and foreign proteins, and some normal proteins. ATP is required for the hydrolysis of proteins, but not peptides by protease La. ATP must bind to the protease La before protein – peptide bond hydrolysis can take place. ATP hydrolysis and the binding of more ATP is necessary for further peptide bond hydrolysis.

Protease La plays an important role in regulation of the 'SOS response', of *E. coli* to DNA damaging agents such as UV light. As part of the 'SOS response' *E. coli* synthesizes the unstable protein SulA which inhibits cell division and allows DNA repair to take place. SulA is stabilized in *lon*$^-$ mutants implicating protease La in SulA degradation. *E. coli* contain an additional ATP-dependent large molecular weight protease activity, protease Clp or Ti (32). Clp consists of two subunit types, one ATP-regulatory which assemble into a 370 kDa oligomer. This protease is expressed constitutively at similar levels, regardless of growth conditions, is not a heat-shock protein, and its role is unclear.

The degradative machinery of bacteria, especially *E. coli* is relevant to the development of biotechnology, as the intracellular fate of foreign proteins of industrial or pharmaceutical importance expressed in these organisms will be determined in part by the degradative activities of the host cell.

5. Proteolytic systems in plants

The bulk of plant proteins appear to be very long-lived or even stable in growing leaves ($t_{1/2}$ of 5 – 8 days), but nevertheless exhibit a variety of half-lives, with regulatory enzymes being particularly short-lived. Relatively little is known about proteolytic systems in plants, but the ubiquitin-mediated system has been detected and the vacuoles of somatic leaf cells and the protein bodies of cotyledons are now thought to be the degradative organelles in plants, analogous to the animal cell lysosome. Many of the known intracellular proteolytic events in plants are analogous to events in animal cells (for example, degradation of abnormal proteins, proteins synthesized in stoichiometric excess, etc.). Unlike animal cells, however, plants utilize storage proteins present in seeds which are mobilized by proteolysis during germination.

The ubiquitin system in plants was first observed in oat seedlings where it is responsible for the rapid degradation of the far-red-light absorbing form of phytochrome (see Section 3.2). Ubiquitin gene expression is increased following a variety of stresses to plants, including heat, therefore ubiquitin-mediated proteolysis may remove stress damaged proteins as in animal cells (33). Ubiquitin-mediated proteolysis is implicated in the loss of organelles and proteins, from cells destined to form the xylem. Perturbation of the ubiquitin system in plants expressing a mutant ubiquitin which cannot form multi-ubiquitin chains (lysine 48 substituted by a cysteine) leads to leaf curling and disruption of vascular tissue formation, suggesting that ubiquitin-mediated proteolysis is part of the remodelling of differentiation (34). Other mechanisms may be responsible for the large-scale proteolysis which occurs in senescing leaves. RuBisCO, probably the most abundant protein on Earth, and other leaf proteins, are degraded in senescing leaves and the amino acids released transported to areas of active growth.

Plant proteins are exposed to particular hazards such as light-induced damage to photosystems which can lead to the generation of abnormal proteins with the potential to disrupt the photosynthetic process. The chloroplast protein D_1 is rapidly degraded following light-induced damage to photosystem II. As with animal cells, plant-cell protein degradation increases during starvation to release amino acids for essential protein synthesis. The lysosomal system is responsible for the starvation-enhanced proteolysis in animals cells, but it is not clear if this is the mechanism responsible in plants.

6. Further reading

Lysosomal proteolysis

Glaumann, H. and Ballard, F. J. (ed.) (1987) *Lysosomes: their role in protein breakdown.* Academic Press, London and New York.

Holtzman, E. (1989) *Lysosomes.* Plenum Press, New York.

Non-lysosomal proteolysis

Orlowski, M. (1990) *Biochemistry,* **29,** 10289. Review article on the multicatalytic proteinase complex.

Rechsteiner, M. C. (1988) *Ubiquitin.* Plenum Press, New York.

Vierstra, R. D. (1989) Protein degradation. In *Biochemistry of plants,* Vol. 15, pp. 521–36. Academic Press, San Diego.

Jentsch, S., Seufert, W., and Hauser, H.-P. (1991) *Biochem. Biophys. Acta,* **1089,** 127. Review article: Genetic analysis of the ubiquitin system.

7. References

1. Plomp, P.J., Gordon, P.B., Meijer, A.J., Hoyvik, H., and Seglen, P.O. (1989) *J. Biol. Chem.*, **264,** 6699.
2. Dunn, W.A. (1990) *J. Cell Biol.*, **110,** 1923.
3. Knecht, E., Martinez, R.A., and Grisolia, S. (1988) *J. Histochem. Cytochem.*, **36,** 1433.
4. Marzella, L. and Glaumann, H. (1987) In *Lysosomes, their role in protein breakdown* (ed. H. Glaumann and F.J. Ballard), pp. 319–67. Academic Press, London.
5. Gordon, P.B. and Seglen, P.O. (1988) *Biochem. Biophys. Res. Commun.*, **151,** 40.
6. Schnell, A.H., Swenne, I., and Borg, L.A. (1988) *Cell Tissue Res.*, **252,** 9.
7. Berman, A.E. and Morozevich, G.E. (1990) *FEBS Letts.*, **263,** 285.
8. Wilkinson, K.D., Urban, M.K., and Haas, A.L. (1980) *J. Biol. Chem.*, **255,** 7529.
9. Rapoport, S., Dubiel, W., and Muller, M. (1985) *FEBS Letts.*, **180,** 249.
10. Chau, V., Tobias, J.W., Bachmair, A., Marriott, D., Ecker, D.J., Gonda, D.K., and Varshavsky, A. (1989) *Science*, **243,** 1576.
11. Bachmair, A. and Varshavsky, A. (1989) *Cell*, **56,** 1019.
12. Jentsch, S., Seufert, W., Sommer, T., and Reins, H.A. (1990) *Trends Biochem. Sci.*, **15,** 195.
13. Redman, K.L. and Rechsteiner, M. (1988) *J. Biol. Chem.*, **263,** 4926.
14. Laszlo, L., Doherty, F.J., Osborn, N.U., and Mayer, R.J. (1990) *FEBS Letts.*, **261,** 365.
15. Gropper, R., Brandt, R.A., Elias, S., Bearer, C.F., Mayer, A., Schwartz, A.L., and Ciechanover, A. (1991) *J. Biol. Chem.*, **266,** 3602.
16. Glotzer, M., Murray, A.W., and Kirschner, M.W. (1991) *Nature*, **349,** 132.
17. Jabben, M., Shanklin, J., Vierstra, R.D. (1989) *J. Biol. Chem.*, **264,** 4998.
18. Parag, H.A., Raboy, B., and Kulka, R.G. (1987) *EMBO J.*, **6,** 55.
19. Melloni, E. and Pontremoli, S. (1989) *Trends Neurosci.*, **12,** 438.
20. Pontremoli, S.S., Melloni, E., Salamino, F., Patrone, M., Michetti, M., and Horecker, B.L. (1989) *Biochem. Biophys. Res. Commun.*, **160,** 737.
21. Denny, J.B., Polancurtain, J., Ghuman, A., Wayner, M.J., and Armstrong, D.L. (1990) *Brain Res.*, **534,** 317.
22. Dahlmann, B., Kopp, F., Kuehn, L., Niedel, B., Pfeifer, G., Hegerl, R., and Baumeister, W. (1989) *FEBS Letts.*, **251,** 125.
23. Rivett, A.J. (1989) *J. Biol. Chem.*, **264,** 12215.
24. Driscoll, J. and Goldberg, A.L. (1990) *J. Biol. Chem.*, **265,** 4789.
25. Eytan, E., Ganoth, D., Armon, Y., and Hershko, A. (1989) *Proc. Natl. Acad. Sci., USA.*, **86,** 7751.
26. Goldberg, A.L., Voellmy, R., Chung, C.H., Menon, A.S., Desautels, M., Meicsell, T., and Waxman, L. (1985) In *Intracellular protein catabolism* (ed. E.A. Khairallah, J. S., Bond, and J. W. C. Bird), p. 33. Alan R. Liss, New York.
27. Schmidt, G.W. and Mishkind, M.L. (1983) *Proc. Natl. Acad. Sci., USA*, **80,** 2632.
28. Bonifacino, J.S., Suzuki, C.K., Lippincottschwartz, J., Weissman, A.M., and Klausner, R.D. (1989) *J. Cell Biol.*, **109,** 73.
29. Klausner, R.D., and Sitia, R. (1990) *Cell*, **62,** 611.
30. Edmunds, T. and Goldberg, A.L. (1986) *Nature*, **322,** 209.
31. Gottesman, S., Clark, W.P., and Maurizi, M.R. (1990) *J. Biol. Chem.*, **265,** 788.
32. Shimogawara, K. and Muto, S. (1989) *Plant Cell Physiol.*, **30,** 9.
33. Bachmair, A., Becker, F., Masterson, R.V., and Schell, J. (1990) *EMBO J.*, **9,** 454.
34. Mayer, R.J., Arnold, J., Laszlo, L., Landon, M., and Lowe, J. (1991) *Biochim. Biophys. Acta*, **1089,** 141.

4

Molecular recognition and intracellular proteolysis

1. Selective and non-selective proteolysis

Proteins that are constitutively unstable must contain signals which consign them to rapid turnover. Recent years have seen the identification of some of the molecular signals residing in proteins that lead to their rapid recognition and degradation by non-lysosomal mechanisms. Lysosomal sequestration of cytoplasmic proteins, on the other hand, has been traditionally viewed as non-selective. It is also often stated that short-lived proteins are degraded by non-lysosomal proteases and that long-lived proteins are degraded lysosomally. However, recent data on the degradation of the short-lived protein fructose 1,6-bisphosphatase in yeast has shown that some short-lived proteins may be degraded in lysosomes. In addition, there is at least one route of lysosomal sequestration that is activated under conditions of nutrient and serum withdrawal which selects cytoplasmic proteins containing a certain pentapeptide motif. This section describes the known molecular signals which target intracellular proteins for degradation.

2. 'Abnormal' proteins

Once in place in their functional cell compartment (cytosol, mitochondrion, etc.), completely synthesized, folded and assembled proteins will be subject to turnover by macroautophagy which proceeds at less than 1% of total cell protein per hour under basal conditions. Short-lived regulatory proteins will be selectively degraded by other mechanisms. It is generally accepted, however, that in addition there are intracellular mechanisms to identify and rapidly degrade proteins that are somehow 'abnormal'. Cells incubated in culture with amino acid analogues (for example, the arginine analogue canavanine) will incorporate the abnormal amino acid into proteins and it is assumed such analogue-containing proteins are abnormal. Indeed, proteins containing amino acid analogues are rapidly degraded with half lives of a few minutes. Prematurely truncated polypeptides, produced by the incorporation of puromycin, are similarly rapidly degraded. Experiments of this kind suggest that a 'proof reading' mechanism(s) exists to degrade mistranslated polypeptides rapidly.

Proteins which fold incorrectly may expose protease sensitive sites, or other signals for their rapid degradation. Following synthesis, polypeptides with N-terminal signal sequences are maintained in an unfolded state by interaction with members of the 70 kDa heat-shock family and directed to non-cytosolic cell compartments. Premature folding of protein precursors in the cytosolic compartments may prevent their translocation and import. The major role of a recently discovered targeting mechanism for ubiquitin-mediated proteolysis (the 'N-end rule', Chapter 4, Section 3) may be for recognition and rapid degradation of just such miscompartmentalized polypeptides. Once folded and in the correct cell compartment, polypeptide subunits find their partners and assemble into oligomers. In the absence of suitable partners, unassembled subunits in chloroplasts, mitochondria, and the endoplasmic reticulum are found to be rapidly degraded. Therefore, subunits synthesized in stoichiometric excess, are judged abnormal.

An abnormal protein structure may also be generated by post-translational mechanisms. Glycoproteins may be incorrectly glycosylated. Proteins, with sulfydryl (cysteine) groups in the reducing environment of the cell cytoplasm, may adopt abnormal conformations, if that reducing environment is compromised by exposure to oxidizing agents. Protein amino acid residues may be modified by the action of metal-catalysed oxidation systems which generate H_2O_2, and via formation of the hydroxyl radical ($\dot{O}H$) modify some amino acid residues by, for example, generating carbonyl derivatives (1). Modification of amino acid side-chains in turn may alter a protein's conformation and signal its rapid degradation. Indeed, glutamine synthetase is found to be rapidly degraded following limited free radical attack on the protein. Therefore, even proteins turned over constitutively at a slow rate in the mature form, may be subject to rapid elimination if they are subject to errors in the translational and post-translational mechanisms, or if the mature protein becomes damaged in some way. The following sections outline what is known about the targeting mechanisms which may allow the cell to identify 'normal' proteins which must be turned over rapidly.

3. The 'N-end rule'

While initially all proteins have methionine at their N-terminal, subsequent processing of the protein can lead to any of the amino acids being present at the N-terminus. The identity of the N-terminal residue of certain proteins may determine whether that protein is a good acceptor for ubiquitin and consequently rapidly degraded (2). The half-lives of the various β-galactosidase test proteins expressed in yeast (*Figure 4.1*) fall into rapidly degraded ($t_{1/2} > 1$ h) and stable groups ($t_{1/2} > 1$ h) which are distinguishable by the amino acid present at the amino-terminus (*Table 4.1*). Basic (Type I substrates, *Table 4.1*) and bulky hydrophobic amino acids (Type II, *Table 4.1*) confer short half-lives when expressed at the amino-terminus. Those β-galactosidase test proteins which are rapidly degraded also readily form ubiquitin conjugates, while proteolytically stable species are poor substrates for ubiquitination. This suggests that the test proteins are selected for ubiquitination and degradation on the basis of their amino-terminal residue. The identity of the amino-terminal residue also determines the degradative fate of β-galactosidase test proteins in reticulocyte lysates (3) (*Table 4.1*).

The acidic amino acids aspartate and glutamate are destabilizing because the proteins bearing these residues at the N-terminus are modified by the N-terminal addition of arginine, a Type I destabilizing residue, by an arginyl-tRNA: protein arginyl transferase. Glutamine and asparagine also promote N-terminal arginylation after deamination to glutamate and aspartate respectively.

For any protein to be ubiquitinated it must contain internal lysine residues which provide the ϵ-amino group for the isopeptide bond to ubiquitin. The N-terminal α-amino group does not post-translationally accept ubiquitin. Therefore, regardless of the signal for recognition, an acceptor lysine(s) is a prerequisite for protein ubiquitination. Not all lysine residues in a protein are available for ubiquitination following amino terminal recognition of a target protein

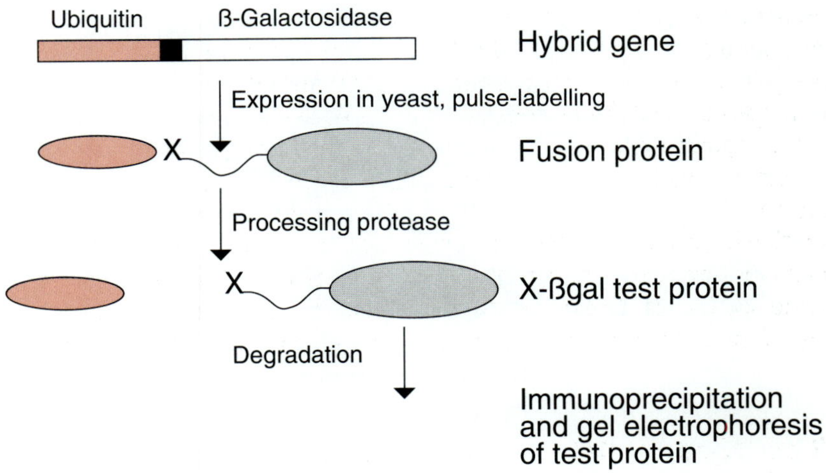

Figure 4.1. Determination of the 'N-end rule'. Expression of hybrid genes encoding a ubiquitin-β-galactosidase fusion protein in yeast is followed by rapid cleavage by a ubiquitin C-terminal processing protease to yield free ubiquitin and β-galactosidase. This processing is exploited together with site-directed mutagenesis at position X to generate variants in the test proteins expressing different N-terminal residues. Degradation of the test proteins in yeast is followed by pulse-labelling and immunoprecipitation of the test protein during the subsequent chase.

Table 4.1 The N-end rule for ubiquitin-mediated proteolysis. Test proteins bearing destabilizing residues at the N-terminus are ubiquitinated and rapidly degraded in yeast and reticulocyte lysate

	N-terminal residue		N-terminal residue
Primary destabilizing residues		*Secondary destabilizing residues*	
Type I (basic)	Arg Lys His		Asp Glu Cys
Type II (bulky hydrophobic)	Phe Leu	*Tertiary destabilizing residues*	
	Trp Tyr Ile†		Asn Gln
Type III (small neutral)	Ala* Ser* Thr*		

*Residues which are destabilizing in reticulocyte lysate but stabilizing in yeast.
†Residues which are destabilizing in yeast and stabilizing in reticulocyte lysate.

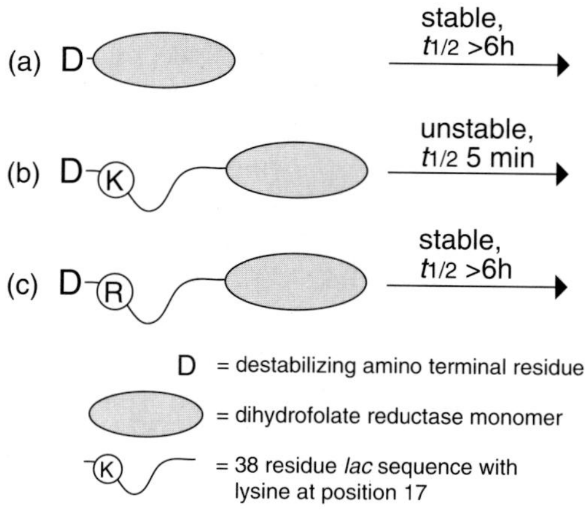

Figure 4.2. A suitable acceptor lysine is required for ubiquitination by the N-end rule. Variants of the dihydrofolate reductase (DHFR) test protein are expressed in yeast. DHFR test proteins bearing N-terminal destabilizing residues are stable (a) unless the 38 residue segmentally mobile *lac* sequence, which contains a lysine at position 17, is present at the N-terminus (b). Substitution of lysine 17 with an arginine (R) stabilizes the test protein (c).

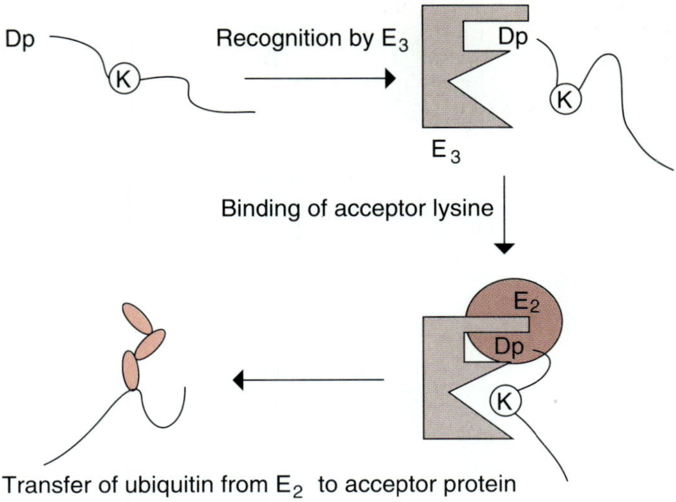

Figure 4.3. Amino-terminal recognition and ubiquitination of proteins. E_3 binds to destabilizing N-termini. A lysine capable of binding to a second site on the E_3 + E_2 complex forms a E_2 catalysed peptide bond with the C-terminus of ubiquitin. Dp is a destabilizing amino terminus.

(*Figure 4.2*). This can be explained if the protein factor (E3) which binds potential N-end-rule substrates for ubiquitination first recognizes and then binds the amino terminus. Binding of an acceptor lysine by the same protein, or by an associated ubiquitin carrier protein (E_2; *Figure 4.3*), is the second step and requires the presence of a lysine residue in a segment of the substrate molecule capable of binding to the appropriate site for ubiquitination to occur.

β-Galactosidase is a tetrameric protein with identical subunits but many oligomeric proteins consist of non-identical subunits. The two determinants of the N-end rule can reside on different subunits and still target a polypeptide for degradation (5). Experiments with heterotetramers of β-galactosidase (*Figure 4.4*) suggest a mechanism whereby the destabilizing amino terminus is bound by the recognition system but ubiquitination can occur on the acceptor lysine of the subunit carrying a stabilizing N-terminal residue (*Figure 4.4*). Therefore, the acceptor lysine can be recognized in either *cis*, or *trans*. The implications of these findings are that heterologous subunits in an oligomeric protein may be degraded at the same or different rates. For differential turnover of subunits to take place dissociation of the oligomer must precede, or be coupled to, degradation (*Figure 4.4*).

A yeast mutant which is unable to degrade the unstable test protein Arg-β-galactosidase shows only subtle differences in phenotype from wild-type cells. Although β-galactosidase proteins with destabilizing amino terminii are stabilized in these mutants, endogenous short-lived yeast proteins are not stabilized (6). Therefore, elimination of the amino-terminal recognizing machinery for protein ubiquitination does not impair the yeast cells' ability to degrade endogenous short-

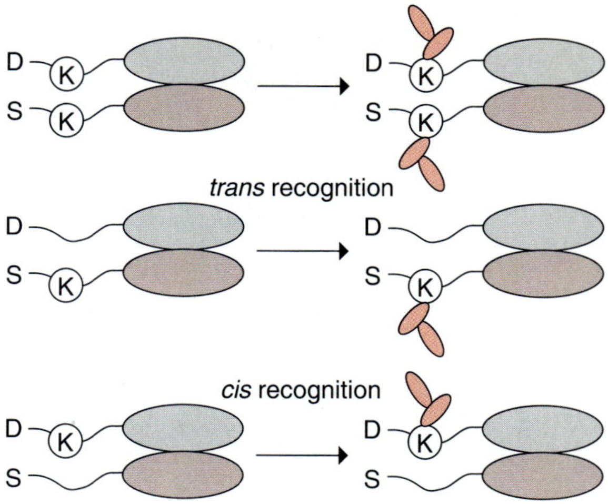

Figure 4.4. *Cis* and *trans* recognition of subunits by the N-end rule. Radiolabelled and non-radiolabelled β-galactosidase test proteins homotetramers are dissociated to monomers in denaturant and radiolabelled monomers mixed with a tenfold excess of unlabelled monomers, differing in the N-terminal residue or the presence of an acceptor lysine. Reassociation by dilution of the denaturant generates heterotetramers with an average of one labelled subunit. Subunits bearing an acceptor lysine but a stabilizing N-terminal residue can be recognized in *trans* when present in heterotetramers containing destabilizing residue-bearing subunits.

lived or 'abnormal' proteins. The great majority of non-compartmentalized proteins (i.e. intracellular and cytosolic) whose sequences are known have N-end-rule stabilizing residues at the amino terminus. In addition, it appears that in the majority of mature cytosolic proteins amino-terminii are blocked by *N*-acetylation. Secretory and compartmentalized proteins, however, more often express destabilizing amino-terminal residues. Possibly the N-end-rule recognition machinery (recognin E_3) is concerned with the rapid removal of miscompartmentalized proteins; for example, secretory proteins which have entered the cytosolic space in error. It must be stressed that the majority of proteins may be recognized by the ubiquitination system by signals other than the N-end rule. Indeed, ubiquitin can be conjugated to proteins in the absence of E_3. Such pathways might recognize hydrophobic surfaces exposed to the aqueous space in damaged proteins (for example, during heat shock) or other targeting sequences in short-lived proteins.

4. PEST and rapid proteolysis

With the accelerating accumulation of protein sequence data held in computerized databases, it has become possible to search for amino-acid sequences common to proteins known to have short half-lives. Comparison of the sequences of ten proteins with intracellular half-lives less than 5 h revealed the presence of regions rich in the amino acids proline (P), glutamate (E), serine (S), and threonine (T) (7). These so-called PEST regions which also contain concentrations of negatively charged residues are commonly flanked by clusters of basic amino acids. PEST regions are highly hydrophilic and would be expected to be on the surface of proteins.

When the sequences of 35 long-lived proteins (half-lives 20 – 200 h) were examined, only three contained recognizable PEST regions. It was possible to predict successfully that the caseins would be rapidly degraded following radiolabelling and microinjection into recipient culture cells on the basis of the presence of PEST regions in these proteins. Ornithine decarboxylase (ODC), a key enzyme in polyamine synthesis, has an intracellular half-life of less than an hour. Mouse ODC contains a PEST region towards the C-terminus and another internal PEST region. Truncation of mouse ODC to remove the C-terminal PEST stabilizes ODC (8). In addition, the ODC of trypanosomes which lacks the C-terminal PEST is stable, but a chimeric protein consisting of the trypanosome sequence, with the addition of the C-terminal PEST containing 36 residues from mouse ODC, is unstable. However, while PEST regions may be identified in short-lived proteins they may not be necessary for rapid degradation. The adenovirus gene product E1A which contains several PEST regions is rapidly degraded in cells expressing the protein, but deletion of the PEST regions does not stabilize the protein.

5. KFERQ and 'molecular chaperones'

Lysosomal degradation and sequestration of cytoplasmic proteins are modulated by a variety of factors, including the supply of nutrients, particularly amino acids and growth factors. Cells in culture show enhanced degradation of cytosolic proteins when grown in reduced (1% v/v) concentrations of serum. Red-blood-cell mediated microinjected experiments have demonstrated that while the lysosomal degradation of some microinjected proteins is regulated by serum, the degradation of others is not, therefore serum withdrawal enhanced lysosomal degradation is, at least in part, selective (9).

Ribonuclease A is one test protein whose lysosomal uptake and degradation is enhanced by serum withdrawal ($t_{1/2}$ decreases from 100 h to 50 h). Cleavage of RNase A with subtilisin yields two fragments, a large protein fragment (RNase S protein) and a N-terminal 20 residue peptide (RNase S peptide). While both fragments are degraded lysosomally following microinjection, only the degradation of the peptide fragment is increased following serum withdrawal. The ribonuclease S peptide can restore serum sensitivity to RNase degradation when recombined with the S protein and confer sensitivity to serum withdrawal when chemically coupled to serum insensitive proteins.

The essential region of the S-peptide for conferring enhanced degradation during serum withdrawal consists of residues 7 – 11 of the peptide, lysine –

phenylalanine – glutamate – arginine – glutamine (Lys – Phe – Glu – Arg – Gln in the three-letter and KFERQ in the single-letter amino acid code). Proteins containing this motif are selectively destroyed in tissues that atrophy during fasting (for example, the liver and kidney).

The KFERQ motif is only found in the RNase family, therefore KFERQ must match a motif that exists in a wide range of intracellular proteins. A search of the computer databases for sequences of proteins whose degradation characteristics in response to serum are known, permitted the elucidation of the simplest compositional motif that would include KFERQ and could predict the response of proteins to serum withdrawal. This motif is best represented by an essential glutamine, preceded by a tetrapeptide sequence comprising a basic residue towards the N terminus (Arg, Lys), followed by a hydrophobic residue (Phe, Ile, Leu, Val), an acidic residue (Asp, Glu), and either a basic, or a hydrophobic residue (Arg, Lys, Phe, Ile, Leu, Val) or the mirror image of this.

A protein homologous to the 70 kDa heat-shock protein family binds to RNase S-peptide and to the KFERQ peptide and promotes lysosomal degradation of RNase and RNase S-peptide in a cell free system. The level of this protein, prp73, increases following serum deprivation. The heat-shock proteins have been called 'molecular chaperones' due to their roles in assisting protein translocation. Members of the 70 kDa heat shock family are involved in the import of proteins into mitochondria and endoplasmic reticulum under normal (non-heat-shock) conditions. The precise role of prp73 in selective lysosomal sequestration remains to be determined. In a comparable way to mitochondrial and endoplasmic reticulum precursor protein import, it may assist in polypeptide unfolding allowing a protein to pass through a lysosomal membrane pore. This model is different from microautophagy and may be an additional mechanism for lysosomal sequestration of cytoplasmic proteins.

Serum and nutrient deprivation of cells also leads to a rapid increase in non-selective macroautophagy which is activated within a few minutes. Prolonged macroautophagy would be expected to lead to a loss of essential proteins. The selective increased uptake of proteins described above is activated upon prolonged fasting. This may help to conserve essential proteins, while enabling the cell to recycle amino acids for the synthesis of key proteins during prolonged nutrient stress.

6. Other signals controlling protein degradation

6.1 The tetrapeptide sequence of adenovirus E1A protein

The adenovirus E1A protein is rapidly degraded in cells expressing this protein, even when its PEST sequences are deleted. Expression of mutant forms of E1A in cells identified the N-terminal four residues of E1A, Met – Arg – His – Ile, as necessary for rapid degradation. Even conservative substitution of any of these amino acids (for example, Arg for His) abolishes the signal for rapid degradation and the tetrapeptide sequence must reside no more than six residues from the

N-terminus to be effective (10). The N-terminal methionine is not removed from E1A following synthesis, so the protein does not possess a destabilizing signal for the N-end rule for ubiquitin-mediated degradation. Bacterially synthesized E1A is stable when microinjected into oocytes, in contrast to the protein synthesized from injected mRNA, suggesting that recognition of the N-terminus occurs as the nascent polypeptide emerges from the ribosome. Binding of some recognition factor to the nascent polypeptide may confer instability to the mature protein.

6.2 The yeast α repressor

Metabolic instability is a common feature of regulatory proteins, including those transiently expressed. The yeast α2 repressor protein is involved in the switching of yeast cell type (**a**, α) prior to mating to form diploid cells. The repressor is extremely short-lived, with a half-life of only 5 min. Dissection of the degradation signals in the repressor was accomplished by expression of full-length and truncated repressor-β-galactosidase fusion proteins in yeast cells (11). β-Galactosidase is destabilized when fused to the full-length repressor. Truncation of the repressor moiety reveals that the repressor contains an amino-terminal and a carboxy-terminal region, independently conferring instability. Genetic analysis suggest that separate proteolytic systems recognize these destabilizing regions.

β-Galactosidase is a tetrameric protein so that heterotetramers can be formed in host yeast cells expressing two different repressor-β-galactosidase fusion proteins. If an unstable fusion protein containing the repressor degradation signal is incorporated into a heterotetramer where the majority of the fusion protein subunits are stable (i.e. lack the repressor degradation signal), the unstable subunit retains its instability, while the subunits which lacked the amino terminal signal remained stable. This implies that the degradative machinery can distinguish between unstable and stable subunits in the same oligomer. Ubiquitinated forms of the repressor have been found *in vivo*, but neither the proteolytic mechanisms nor the precise degradative signals in α-repressor degradation have yet been clearly identified.

6.3 Cyclin and the 'destruction box'

The ubiquitin-mediated degradation of the cell-cycle regulator cyclin is necessary for transition from metaphase to anaphase (Chapter 3, Section 3.2). Expression of mutant cyclins has established that there are two regions necessary for rapid degradation of cyclin (12). Residues 13–66 contain a nine amino-acid sequence (single amino-acid code: RAALGNISN), the 'destruction box' which is necessary for rapid degradation, and a region rich in lysines which may serve as a ubiquitination site and, indeed, the degradation of cyclin appears to be ubiquitin-mediated. As the degradation signal is internal, this protein would appear not to be selected by the N-end rule. Interestingly, a region of weak homology to the cyclin destruction box is found in a region of the yeast α2-repressor essential for its rapid degradation.

7. A hierarchy of molecular recognition signals for protein degradation

It should be clear from the previous sections that there are a number of molecular recognition signals for intracellular protein degradation which have been identified, and no doubt more will be found. Some degradation signals may be recognized only under special conditions (for example, serum deprivation) and some proteins may be destabilized following association with other proteins. Association of the papillomavirus oncoprotein E6 with the cellular transformation suppressor p53 results in rapid degradation of p53 by the ubiquitin system (a possible example of recognition in *trans*) and cell transformation. Finally, expression of a degradative signal may require post-translational modifications, such as N-terminal modification, or exposure of previously hidden destabilizing regions following denaturation events (cleavage of disulfide bonds, oxidation of amino acid side-chains, partial proteolysis). Long-lived proteins may well lack any protein degradation signal, their turnover being dependent on the constitutive but inducible process of macroautophagy.

The difficulty in identifying intermediate products of protein catabolism in the cytoplasm of cells suggests that once a protein has been identified for degradation a single cytoplasmic proteolytic system is capable of degrading it to small peptides and amino acids. Alternatively, that following the initial proteolytic events, the recruitment of different proteases, perhaps by exposure of new destabilizing features, ensures rapid and complete catabolism of the proteins.

8. Further reading

Dice, J.F. (1990) *Trends Biochem. Sci.*, **15**, 305. Review article: Peptide sequences that target cytosolic proteins for lysosomal proteolysis.

Hershko, A. (1988) *J. Biol. Chem.*, **263**, 15237. Review article: Ubiquitin-mediated protein degradation.

Rechsteiner, M.C. (1988) *Ubiquitin*. Plenum Press, New York.

9. References

1. Stadtman, E.R. and Oliver, C.N. (1991) *J. Biol. Chem.*, **266**, 2005.
2. Bachmair, A., Finley, D., and Varshavsky, A. (1986) *Science*, **234**, 179.
3. Gonda, D.K., Bachmair, A., Wunning, I., Tobias, J.W., Lane, W.S., and Varshavsky, A. (1989) *J. Biol. Chem.*, **264**, 16700.
4. Bachmair, A. and Varshavsky, A. (1989) *Cell*, **56**, 1019.
5. Johnson, E.S., Gonda, D.K., and Varshavsky, A. (1990) *Nature*, **346**, 287.
6. Bartel, B., Wunning, I., and Varshavsky, A. (1990) *EMBO J.*, **9**, 3179.
7. Rogers, S. and Rechsteiner, M. (1987) *Science*, **234**, 364.
8. Ghoda, L., Wetters, T.V., McCrae, M., Ascherman, D., and Coffino, P. (1989) *Science*, **243**, 1493.
9. Chiang, H.-L. and Dice, J.F. (1988) *J. Biol. Chem.*, **263**, 6797.
10. Simon, R. and Richter, J.D. (1990) *Mol. Cell. Biol.*, **10**, 5609.
11. Hochstrasser, M. and Varshavsky, A. (1990) *Cell*, **61**, 697.
12. Glotzer, M., Murray, A.W., and Kirschner, M.W. (1991) *Nature*, **349**, 132.

5

Regulation of intracellular protein turnover

Different methods of regulating protein degradation occur in different types of cells. Since proteins are continuously synthesized and degraded in most cells, the two processes must be co-ordinated in some way. The relationship between protein synthesis and degradation might be expected to be different in different tissues and this is indeed found to be the case. Protein synthesis rates may be relatively constant in some tissues and vary more dramatically in others in changing metabolic conditions. Alternatively, protein degradation rates may be relatively constant, or vary. Examples of both types of regulation of protein turnover will be described. The regulation of the turnover of two enzymes, HMG-CoA reductase and fructose 1,6 bisphosphatase, will be presented as examples of the complex regulation of the turnover of key regulatory proteins.

1. Regulation of liver protein degradation

The liver is one site of 'storage' of dietary amino acids. There is no 'store' of proteins, unlike triglyceride and glycogen for lipids and glucose, respectively. Functional intracellular proteins are synthesized and degraded continuously. Liver protein concentration can decrease at the rate of 2% per hour during starvation and increase at the same rate on refeeding. The rate of synthesis of liver proteins remains relatively constant during starvation, with the rate increasing somewhat on refeeding. In contrast, the rate of protein degradation can change enormously varying from 70% per day during starvation to 11% per day on refeeding. The rate of protein degradation controls the protein content of liver and must be finely controlled so that it can alter in response to the dietary changes.

The degradation of liver proteins in starvation is brought about by macroautophagy (Chapter 3, Section 2.1). The volume fraction of autophagosomes (as a percentage of cell organelles) can increase within minutes when a perfused liver is deprived of amino acids to reach a maximum after 20 min and remain constant at this level for up to 90 min. The objective of increased degradation in starvation is to provide amino acids for gluconeogenesis and other metabolic processes. In perfused liver, the addition of amino acids can stop the formation of autophagosomes instantaneously. Approximately half of the twenty amino acids have an inhibitory role on macroautography in liver. The amino acids which inhibit, some of which are non-essential, are leucine, phenylalanine, tyrosine, tryptophan, histidine, methionine, asparagine, glutamine, and proline. The effects of the amino acids are additive. Asparagine and glutamine are particularly effective and may act by inhibiting lysosome-autophagosome fusion. Leucine is also particularly effective in the inhibition of degradation and may contribute up to half of the inhibitory capacity of the amino acids in perfused liver.

The most likely mode of action of amino acids is to prevent the formation of autophagosomes, thereby inhibiting macroautophagy. The biochemical mechanisms by which amino acids can perform this function are unknown, but the effect is very rapid and persistent. In animals refed for 12 h, macroautophagy is reduced by up to 90%. It is not known if the amino acids themselves, or some metabolites of amino acids, are the potent regulators of autophagy. The formation of autophagosomes is dependent on the wrapping of membrane around target cytoplasm. The membrane may be derived from the endoplasmic reticulum, or may be derived from a specific type of membrane organelle in hepatocytes called a 'phagophore'. The wrapping process requires ATP and presumably can be regulated by the usual protein kinase modulatory systems. The formation of autophagosomes is also very temperature-dependent and may require some signal, perhaps a nucleation event in the cytoplasm (for example, ubiquitination, Chapter 3), as well as the mechanism to sequester cytoplasm within a membrane.

Prolonged stress also induces a selective mechanism of lysosomal sequestration (Chapter 4, Section 5). The cytosolic proteins which are selectively degraded in serum-deprived stressed cells may have evolved for selection for degradation during stress-induced proteolysis. These proteins may be dispensable proteins which can be degraded to provide amino acids for the synthesis of proteins essential for cell survival. Therefore, although 'storage' proteins are not generally thought to exist in cells, some proteins are more 'dispensable' than others.

The rate of protein degradation in liver can also be controlled hormonally; for example, glucagon substantially increases liver protein degradation. This process is mediated by the increased formation of autophagosomes which is opposed by the action of insulin. Insulin can cause the size of the autophagolysosome compartment to contract dramatically within minutes (for example, with a $t_{1/2}$ of 8 – 10 min). These rapid changes in the size of the autophagolysosomal compartment means that rapid changes in the rate of protein degradation can occur in liver. This makes the hepatocytes remarkably effective cells for the maintenance of the free nitrogen pool (largely contributed to by amino acids) in the circulation.

1.1 Protein degradation and liver growth

Liver growth occurs developmentally, and in some animals in response to partial hepatectomy. Protein synthesis rates remain remarkably constant in these situations. The increase in liver mass is due to the combination of ***hyperplasia*** (increase in cell numbers) and ***hypertrophy*** (increase in cell size). The latter is caused by decreased rates of hepatocyte protein degradation. The degradation rates of proteins in liver organelles (for example, mitochondria) and individual enzymes are considerably reduced during neonatal development, therefore protein accumulation by the liver in development, like in refeeding or following hepatectomy in some animals, is achieved by decreases in the rate of protein degradation. This is generally accomplished by suppression of macroautophagy.

2. Protein turnover in developing mammary gland

The hormonal development of the mammary gland is accompanied by enhanced specific gene expression for milk proteins and the enzymes which produce the milk fats and lactose. Net casein production is modulated at the protein level in response to changes in suckling by differential lysosomal degradation of casein. The accumulation of the enzymes which make the milk fats (such as fatty acid synthetase) is achieved by hormonal-dependent increases in the rate of synthesis of the enzymes and decreases in the rate of degradation of the enzymes. The control of both protein synthesis and degradation for mammary development is mimicked at the mRNA level, where the enormous increase in the concentration of casein mRNAs is achieved by a modest increase in the rate of gene transcription, coupled with a large increase in the stability of the mRNAs. The mRNA degradation system must be hormonally inactivated in a way which is not presently understood.

3. Regulation of muscle protein degradation

Difficulties in estimating muscle protein turnover, combined with the relatively slow turnover of muscle proteins (4 – 5% per day compared with 40 – 50% per day in liver), has limited experimentation and therefore understanding of the regulation of protein turnover in a variety of physiological and clinical situations.

3.1 Muscle growth and development

The most striking observation of protein metabolism in muscle is that rates of both protein synthesis and protein degradation are much higher in the earlier than in the later stages of development (approximately 30% per day in fetal skeletal muscles down to 4 – 5% per day in adult skeletal muscles). The important extension of this observation is that the accumulation of muscle protein is due to the difference between the rate of protein synthesis and degradation. This type of mechanism in which the change in muscle protein mass is due to the difference in the two rates seems to apply to situations where there is an increase (for example, development) or decrease (for example, atrophy) of muscle protein content. The decrease in protein turnover during development applies to both skeletal and cardiac muscles. Changes in muscle protein content have a dramatic effect on overall nitrogen balance in the body in terms of the pool of free amino acid (for example, circulatory versus the pool of bound amino acid in proteins), since, in the adult, muscles can account for up to 25% of body weight, and it has been estimated that the replacement of the proteins continuously degraded in muscle can consume up to 25% of the ATP production in muscle.

3.2 Changes in muscle protein turnover in mature muscles

Many experimental studies have examined the changes in muscle protein synthesis and degradation in response to such variables as exercise, immobilization, stretch, denervation, and starvation. In general, it appears that protein synthesis in muscles responds much more rapidly to imposed changes than protein degradation, although both processes usually alter to some extent in most situations. For example, if muscle is immobilized in a lengthened state (for example, by limb immobilization in a plaster cast), then muscle growth is achieved by a rapid increase in protein synthesis (in 12 – 24 h) accompanied after a lag (24 – 48 h) by a compensatory increase in protein degradation. In starvation, there is a rapid decrease in skeletal muscle protein synthesis accompanied by a much slower fall in protein degradation. The excess degradation over synthesis results in amino acid production needed to fuel liver gluconeogenesis for survival during prolonged starvation. Clearly, in the long-term, this unstable state will lead to progressive muscle-wasting and loss of function. On refeeding there is a rapid increase in protein synthesis with protein degradation again returning more slowly to normal rates. These types of changes in protein synthesis and degradation are not surprising in muscle when the architectural complexity of the myofibril is considered, synthesis and insertion of new proteins being ostensibly easier relative to the excision and removal of proteins which function in the myofibrillar structure. Addition of newly synthesized proteins to the periphery of the myofibril may make synthesis and biogenesis easier than catabolism, but the overall processes are currently not completely understood. Lysosomes have been implicated in muscle protein breakdown; however, the architectural constraints of myofibres restrict lysosomes to the periphery of the fibres. The manner in which muscle proteins can enter the lysosomal system is, therefore, not yet clear. Perhaps limited proteolysis occurs to generate fragments which can more readily enter the lysosome; for example, mediated by the calcium-dependent calpains. Alternatively, the ubiquitin-dependent proteolytic system and the multicatalytic protease (proteasome) may have a mediatory role.

Insulin and other hormones have a key role in the control of muscle protein turnover. Insulin rapidly stimulates the synthesis of muscle proteins in response to feeding, so consuming a large proportion of amino acids which have entered from the circulation. Insulin may also suppress the degradation of proteins. In heart muscle, insulin causes a decrease in protein degradation which may be mediated through the known inhibitory effect of the transamination products of branched-chain amino acids. The ketoacids probably inhibit amino-acid branched-chain dehydrogenases. This elevates levels of branched-chain amino-acids in muscles which inhibits protein degradation. Insulin is not alone in the control of muscle protein turnover. Glucocorticoids have an opposing role to insulin and increase protein degradation (for example, in starvation when insulin levels are low), to cause the release of the amino acids from muscle necessary for hepatic gluconeogenesis. The hormonal situation is even more complex in the case of anabolic steroids which increase the muscle mass of animals in meat production and man for better athletic performances. The increase in muscle protein mass caused by anabolic steroids (for example, trenbolone acetate) is mainly due to a decrease in protein degradation rate (as in catch-up growth in children and refed starving animals) with the larger rate of synthesis over degradation accounting for the increase in muscle protein mass.

4. Regulation of the degradation of proteins in cultured cells

Generally, anabolic growth factors with pleiotropic effects on cells bearing appropriate receptors, including stimulation of nutrient uptake, DNA synthesis, RNA synthesis, and protein synthesis, cause substantial decreases in rates of protein catabolism in cells deprived of serum. The effects are thought to be produced by a decrease in macroautophagy induced in the cells by the serum deprivation. The removal of serum from cells is often said to induce a 'step-down' increase in proteolysis: it is this macroautophagic increase which is inhibited by anabolic growth factors. The anabolic growth factors include insulin, insulin-like growth factors (IGF-1; IGF-2), and epithelial growth factor (EGF). In almost every cell line with receptors, these factors induce an increase in protein synthesis and a corresponding decrease in protein degradation. Hepatocytes growing in the presence of serum respond to glucagon and adrenalin with increases in protein degradation rates. Other cell types are more reluctant to respond similarly to these agents.

Muscle-derived cells have been shown to respond to glucocorticoids by an increase in protein degradation similar to *in vivo*. Glucocorticoids have been shown to cause an increase in protein catabolism and decrease in protein synthesis. The thyroid hormone induced increases in the rates of protein catabolism in muscle and liver *in vivo* are difficult to reproduce in cultured cells. This is probably because their *in vivo* actions are complex resulting in the stimulation of the release of other hormones and circulatory factors which cause the direct effects seen in target cells.

Studies on protein turnover in cultured cells have provided interesting corroborative information on the effects of ligands *in vivo*. Furthermore, the studies have led to theories of growth control which are important for the understanding of the general effects of growth factors which are increasingly seen to be important homeostatic regulators *in vivo*. Generally, both protein synthesis and degradation are controlled in cultured cells.

5. The turnover of regulatory enzymes

This section describes the regulated protein turnover of two enzymes, hydroxymethylglutaryl coenzyme A reductase (HMG-CoA reductase) and yeast fructose 1,6-bisphosphatase (FBPase), both of which are degraded in a membrane bordered cell compartment. An example of an enzyme degraded in the cytosol in a regulated fashion would be ornithine decarboxylase (ODC) which is described in Section 4 of Chapter 4. HMG-CoA reductase is one of the regulatory enzymes of the mevalonate pathway involved in isoprenoid synthesis. This pathway is complex, with end-products which include cholesterol, steroid hormones, vitamin D, dolichol, haem A, ubiquinone, and farnesylated proteins. The pathway shows acute and chronic control of the regulatory enzymes, including HMG-CoA reductase.

The production of the enzyme is regulated at the transcriptional and post-transcriptional levels. Post-transcriptional regulation includes control of translation of the mRNA and rate of degradation of the enzyme. Like most regulatory proteins the enzyme has a short half-life ($t_{1/2}$ 2 – 4 h). The enzyme is a multiple membrane-spanning transmembrane glycoprotein in the endoplasmic reticulum membrane. Mutagenesis experiments have shown that the degradation rate of the enzyme is controlled by the membrane-spanning domain and is controlled by sterols in a feedback manner. The rate of degradation of the enzyme increases when the concentration of sterol rises in the cell (1). The sterols act to produce this effect by interacting with the membrane-spanning domain of the enzyme. There is also evidence that a non-sterol isoprenoid, as well as a sterol, is necessary for the accelerated degradation of this enzyme. This combination of feedback regulators can accelerate the degradation of the enzyme fivefold. The magnitude of this effect is similar to changes in the rate of translation of the mRNA. The regulation of HMG-CoA reductase needs multiple mechanisms, in view of the complexity of the mevalonate pathway. Regulation of the degradation rate of the enzyme is one of these important control points. The site of degradation of the reductase is not known, but it may occur in the endoplasmic reticulum.

FBPase is a regulatory enzyme of gluconeogenesis whose synthesis in yeast is repressed and degradation accelerated when cells are grown on glucose. The rapid fall in enzyme concentration on exposure to glucose prevents a futile cycle of fructose-6-phosphate phosphorylation and dephosphorylation occurring. In the presence of glucose, FBPase is rapidly imported into the yeast vacuole (2), a membrane-bordered organelle analogous to the animal cell lysosome and degraded. The import of FBPase into the vacuole involves the secretory apparatus. It may be that exposure to glucose results in the expression of an import protein which is incorporated into the vacuolar membrane following its transit through the endoplasmic reticulum and the Golgi. This putative import protein may be responsible for the binding and import of FBPase. The example of yeast FBPase reinforces the point that the lysosome system is not restricted to the non-selective degradation of long-lived proteins.

6. Further reading

Glaumann, H. and Ballard, F.J. (ed.) (1987) *Lysosomes: their role in protein breakdown.* Academic Press, London.

Waterlow, J.C. and Garlick, P.J. (1978). *Protein turnover in mammalian tissues and in the whole body.* North-Holland, Amsterdam.

7. References

1. Brown, M.S. and Goldstein, J.L. (1990) *Nature,* **343**, 425.
2. Chiang, H.L. and Schekman, R. (1991) *Nature,* **350**, 313.

6

Protein degradation and disease

The delicate balance between protein synthesis and degradation in all tissues might be expected to be disturbed in disease and trauma, and this is indeed the case. Pathological changes of different types can, of course, alter the rates of protein synthesis or protein degradation, or both processes in any tissue. Examples of each of these changes exist. The changes in protein turnover in trauma, starvation, and diabetes will be considered since they are reasonably well documented for man and animals. Protein degradation (and synthesis) in disease will be considered under the headings of metabolic change in whole organisms and tissues and molecular cell biological changes in individual cells.

1. Metabolic changes

Protein turnover (synthesis and degradation) changes in a variety of conditions, including trauma (burns, skeletal abnormalities, surgery and sepsis), starvation, and diabetes. The earliest observation on protein catabolism in disease was an increase in urinary nitrogen, but this does not reveal whether the effect arises from whole-body or organ-specific changes in protein synthesis or degradation. Alternative forms of trauma can cause different changes in protein synthesis and degradation. For example, the results of one study on surgery show a decrease in whole-body protein synthesis with little or no change in protein degradation. However, the nutritive consumption of the patients after surgery was lowered. The consequences are an imbalance in protein turnover which results in the loss of proteins from the body. Burns, sepsis, and skeletal trauma can be associated with an increase in whole-body protein degradation and protein synthesis. In this case the nutritive input was maintained, or increased after trauma. The degree of traumatization of the patient contributes to the degree of imbalance between whole-body protein synthesis and degradation. In general, the gap between the rate of protein degradation and protein synthesis increases with the degree of trauma resulting in an increased breakdown of protein in the body.

The biochemical and endocrine reasons for the imbalance between synthesis and degradation of proteins are complex. Post-trauma loss of insulin sensitivity of muscle and adipose tissue probably results in impairment of protein synthesis. This insulin-resistance, coupled with glucocorticoid production, probably increases the rate of protein breakdown, particularly in skeletal muscle. Treatment of patients with post-trauma loss of protein from the body includes administration of insulin in order to combat the decreased insulin sensitivity of tissues and to produce an anti-catabolic effect.

The whole-body increase in protein degradation can be accounted for by different tissue or cell mechanisms depending on the cause of trauma. For example, in the case of burns, increased protein degradation may be mediated by scavenging cells in order to remove damaged proteins, cells, and tissues. In the cases of sepsis and fractures, an elaborate network of responses by inflammatory cells (for example, macrophages) occurs producing regulatory proteins which are secreted to activate the cellular inflammatory response. Some of these factors may enhance protein degradation in tissues and result, directly or indirectly, in the activation of proteases in the body fluids which can cause proteolysis of circulatory proteins.

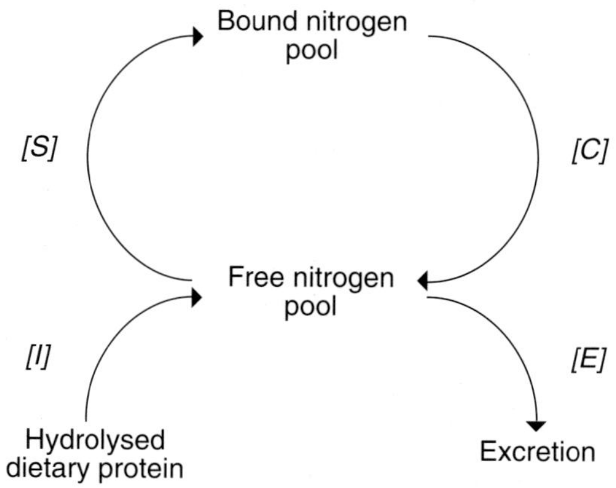

Figure 6.1. Whole-body nitrogen balance. The mobilizable bound nitrogen and free nitrogen in the body is predominately in the form of amino acids but also includes some nucleic acids and other molecules. The bound nitrogen is therefore in the form of proteins. The free form represents the amino acids in the body fluids. The bound pool of nitrogen is increased by protein synthesis [*S*] and decreased by protein degradation [*C*]. The free nitrogen pool can receive amino acids from the catabolism of proteins and also from the degradation of dietary proteins in the gastrointestinal system [*I*]. Nitrogen is removed from the free pool by excretion from the body [*E*]. Normal adults are in nitrogen balance, as indicated in the text. Positive nitrogen balance occurs in growth or during convalescence after illness. Negative nitrogen balance occurs in starvation, debilitating disease, and even by just staying in bed!

Starvation is defined by a lack of dietary nutrients. Normally, the body is in a state of nitrogen balance which is defined in *Figure 6.1*. Nitrogen (mainly amino acids) in the body is found in a free pool in the body fluids or in a bound pool of nitrogen (mainly in the form of proteins). The nitrogen balance equation is given by

$$I + C = S + E. \qquad [6.1]$$

Input into the free pool is from hydrolysed dietary proteins [I] and protein catabolism [C]. Output from the free pool into the bound pool is by protein synthesis [S]. Excretion from the body [E] also removes nitrogen from the free pool. In starvation there is no dietary input [I] so the equation simplifies to

$$C = S + E. \qquad [6.2]$$

There must be modifications in protein turnover and associated processes to maintain the balance between the free and bound pool of nitrogen and to conserve nitrogen in the body. The adaptations include a rapid decrease in protein synthesis, particularly in skeletal muscle, a slow decrease in protein degradation, again particularly in skeletal muscles, a decrease in urea cycle activity and an increase in the reutilization of amino acids (from 75 to 95%). The large reutilization of amino acids in normal conditions (75%) is a measure of the body homeostatic systems, designed to retain nitrogen. In starvation, almost all amino acids are reutilized. The slow decrease in protein degradation is because of the two conflicting needs of the provision of amino acids for the use of their carbon skeletons in gluconeogenesis and the need to conserve muscles, particularly for locomotion. In prolonged starvation, muscle protein degradation always exceeds muscle protein synthesis, although the difference between the two processes declines substantially with time. The consequence is prolonged muscle-wasting and eventual death. This excess degradation over synthesis can be seen in the release of alanine from skeletal muscle, which increases in the first few days of fasting and is reduced after 3 to 6 weeks of starvation. This is mimicked by to some degree, by branched-chain amino acids (leucine, isoleucine, valine) which are elevated in the blood during the first week of starvation and then progressively decline. These changes in protein turnover are mediated by a decline in circulating insulin, together with the protein catabolic effects of circulating glucocorticoids.

Protein loss from the body is also a feature of Type I diabetes. Again, the absence of insulin, together with the presence of circulatory glucocorticoids, will cause muscle protein degradation to exceed protein synthesis and result in protein loss and muscle-wasting.

2. Ageing

The nature of the ageing process remains somewhat mysterious, but many have suggested that the ageing of cells may result from the accumulation of errors in the biosynthesis of proteins leading to compromised cell functions. Abnormal proteins may also be generated by post-translational events, and catalytically inactive forms of some enzymes are found to accumulate in ageing cells and tissues (1). One such inactivation mechanism is the metal-ion catalysed formation of protein carbonyl groups which accumulate in proteins from aged erythrocytes and fibroblasts from old human donors. Carbonyl groups are also more abundant in the proteins of patients with the premature ageing diseases progeria and Werner's syndrome. Proteins damaged by oxidative mechanisms (Chapter 4, Section 2) appear to be readily degraded by large molecular weight cytosolic proteases. The accumulation of enzymatically inactive forms of glutamine synthetase and glucose-6-phosphate dehydrogenase in hepatocytes from old rats suggests that the proteolytic mechanisms for clearing such damaged proteins is impaired by age. Ageing cells may, therefore, become increasingly incapable of clearing damaged proteins as a response to oxidative stress.

While is still remains unclear if general protein turnover declines with age, the lysosomal degradation of some microinjected proteins is found to decline with senescence of human fibroblasts in culture, as does the enhanced lysosomal proteolysis of selected proteins on serum withdrawal (Chapter 4, Section 5) (2). It is likely, therefore, that the lysosomal system is affected in some way by ageing and, indeed, aged brain is often found to accumulate a fluorescent lipoprotein pigment, lipofuscin.

Some cells in the human body are not subjected to cell turnover (for example, neurons) but do turn over their proteins. An exception is the eye lens, where there appears to be no constitutive turnover of cell components and, as a consequence, the lens of the eye is faced with irreplaceable loss of proteins due to oxidative and other stresses. The eye lens appears to possess ubiquitin conjugating and multicatalytic protease (proteasome) activity, with the potential to degrade abnormal proteins. It may be that these and other systems may decline in efficacy with age of the lens, and thus permit the accumulation of aggregated proteins and cataract formation. In conclusion, there are good reasons to believe that the accumulation of abnormal proteins, as a result of a decline in the protective functions served by intracellular proteolytic systems, may be an important part of cell ageing.

3. Ubiquitin in chronic degenerative disease

Several examples of the involvement of the cellular protein degradative systems in chronic degenerative and viral disease have been reported. These may be abberrations in the breakdown of single proteins. Ubiquitin–protein conjugates are present in neurofibrillary tangles of Alzheimer's disease and intracellular inclusions in other major human neurodegenerative diseases; for example, brainstem Lewy bodies in Parkinson's disease and cortical Lewy bodies in diffuse Lewy body disease, as well as in Mallory bodies in alcoholic liver disease and cytoplasmic bodies in cytoplasmic body myopathy (3). Additionally, ubiquitin–protein conjugates are present in filamentous inclusions of anterior horn cells in the spinal cord and in the motor cortex in motor neuron disease (amyotrophic lateral sclerosis). There is also at least a two fold increase in the expression of the heat-shock inducible human polyubiquitin C gene in the brain and spinal cord in the disease. Ubiquitin – protein deposits have also been found in the transmissible human encephalopathies (such as Creutzfeldt – Jakob disease). Therefore, some forms of ubiquitin – protein deposit has been shown to be a feature of both the idiopathic and transmissible encephalopathies. The presence of ubiquitin – protein conjugates may indicate attempts by the cell to remove damaged or abnormal proteins by catabolism which is in some way impaired, resulting in their accumulation in insoluble protein aggregates, along with other cellular components. Interestingly, insoluble intracellular protein inclusions, identical to those seen in Alzheimer's disease (neurofibrillary tangles), are found in the brains of normal old monkeys.

4. Further reading

Beynon, R.J. and Bond, J.S. (ed.) (1989) *Proteolytic enzymes: a practical approach.* IRL Press, Oxford.

5. References

1. Starke-Reed, P.E. and Oliver, C.N. (1989) *Arch. Biochem. Biophys.,* **275,** 559.
2. Dice, J.F. (1989) *Exp. Gerontol.,* **24,** 451.
3. Mayer, R.J., Arnold, J., Laszlo, L., Landon, M., and Lowe, J. (1991) *Biochim. Biophys. Acta,* **1089,** 141.

Glossary

Anabolic: results in build-up of body constituents.

Anterograde: movement of molecules from the site of synthesis in a cell to their site of function.

Atrophy: decrease in size and functional capabilities of a cell, tissue, or organ.

Autophagolysosome: product of the fusion of a lysosome and an autophagosome.

Autophagosome: a membrane-bordered organelle, probably formed from the endoplasmic reticulum which sequesters cytoplasm prior to fusion with a lysosome. Also known as autophagic vacuole.

Autophagy: 'self-eating', the uptake and degradation of cytoplasmic components in the lysosomal system.

Catabolic: results in breakdown of body constituents.

Cathepsin: a lysosomal proteinase.

Crinophagy: intracellular degradation of secretory proteins by fusion of secretory vesicles with lysosomes.

Half-life: time taken for a reaction to go to half-completion; for example, for protein degradation for half of the protein molecules to be degraded.

Heteroautophagy: intracellular degradation of heterologous protein, i.e. uptake and degradation of protein from extracellular space.

Hyperplasia: increase in cell number in a tissue.

Hypertrophy: increase in cell size in a tissue.

Lysosome: membrane-bordered organelle containing hydrolytic enzymes, a major site of the catabolism of intracellular macromolecules, including proteins.

Macroautophagy: bulk sequestration by autophagosomes of cytoplasmic constituents and their lysosomal degradation. Probably non-selective.

Microautophagy: uptake of cytoplasmic proteins into the lysosomal apparatus by invagination of the surface of an organelle of the autophagic/lysosomal system. Possibly selective.

Microinjection: experimental introduction of (macro)molecules into the cytoplasm of cells.

Multivesicular dense body: membrane-bordered organelle, containing small vesicles, possibly derived from the endosomal compartment.

Nitrogen balance: the state where nitrogen intake in diet (e.g. as amino acids) is equalled by excretion of nitrogen (as ammonia and urea).

Pleiotropic: multiple regulatory functions.

Primary lysosome: lysosome which has not engaged in degradation.

Pulse–chase: pulse labelling of a pool of molecules with radiolabelled precursor, removal of radiolabelled tracer, followed by further incubation to follow loss of radiolabelled molecules; for example, pulse-labelling of protein with amino acid, removal of labelled amino acid, and incubation to follow loss of label from protein by degradation.

Pulse-labelling: labelling of a pool of molecules by incorporation of a radioactive tracer over a short period of time; for example, incorporation of radiolabelled amino acid into proteins.

Retrograde: movement of molecules from their site of function in the cell to the site of degradation.

Recognin: component of the ubiquitin-mediated pathway which selects protein substrates; i.e. for the N-end rule E_3.

Secondary lysosome: lysosome which has already been engaged in degradation.

Turnover: synthesis and degradation of biological molecules.

Index